T0133034

Charles Bell and the Anatomy of Reform

Charles Bell and the Anatomy of Reform

CARIN BERKOWITZ

The University of Chicago Press

CHICAGO AND LONDON

Carin Berkowitz is director of the Beckman Center for the History of Chemistry at the Chemical Heritage Foundation. She lives in Swarthmore, Pennsylvania.

The University of Chicago Press, Chicago 60637
The University of Chicago Press, Ltd., London
© 2015 by The University of Chicago
All rights reserved. Published 2015.
Printed in the United States of America

24 23 22 21 20 19 18 17 16 15 1 2 3 4 5

ISBN-13: 978-0-226-28039-4 (cloth)
ISBN-13: 978-0-226-28042-4 (e-book)
DOI: 10.7208/chicago/9780226280424.001.0001

Library of Congress Cataloging-in-Publication Data
Berkowitz, Carin, author.
Charles Bell and the anatomy of reform / Carin Berkowitz.
pages cm
Includes bibliographical references and index.
ISBN 978-0-226-28039-4 (cloth : alk. paper) — ISBN 978-0-226-28042-4 (e-book) 1. Bell, Charles, Sir, 1774–1842. 2. Physicians—England—Biography. 3. Medicine—England—History. I. Title.
R489.B5B47 2015
610.92—dc23
[B]

2014045120

♾ This paper meets the requirements of ANSI/NISO Z39.48-1992 (Permanence of Paper).

For Theo, my greatest gift.

Contents

Introduction

On October 12, 1831, Charles Bell—an anatomist and surgeon, an Edinburgh transplant in London, who was born to a relatively poor family and raised by his mother and his brothers—returned to his London home to find that he had been called to be knighted by the King that morning. By his own account, it was shocking news for which Bell felt hopelessly unprepared: "I found the impossibility of going to Court for reasons which had influenced many a better man before me—the state of my nether garments!"[1] Despite his family's protests and their attempts to dress him in borrowed court dress (attempts that left Bell feeling rather silly), Bell made his excuses for not attending that day, instead deferring until the next levee.

In receiving the knighthood of the Guelphic Order at that time, Charles Bell found himself among luminaries of science, such as the astronomer John Herschel; the mathematician Charles Babbage; David Brewster, whose work broadly spanned the physical sciences; James Ivory (mathematician); and John Leslie (mathematician and physicist); for it had been determined that the Guelphic Order should become the mark of distinction for scientific men. And when they did show up to receive their honor, Bell persuaded Herschel that he "represented the higher sciences," and that Herschel should therefore precede him. While all began as planned, and Herschel did precede Bell into the presence chamber, "in approaching the lord in waiting [Herschel] lost heart, and suddenly countermarched," so that, Bell recounted, "I found myself in front. My niece's dancing-master having acted the king the night before, I had no difficulty."[2] The rehearsal had stood him in good stead for the ceremony of the day. Thus, a man who worried his whole life about money, standing, and a Scottish dialect that forever plagued him, became a knight.

Bell's knighthood shows the high esteem in which he was held as a leading British anatomist and medical man during the tumultuous years of the early nineteenth century, and yet today his name remains little known compared with those of John Herschel or Charles Babbage, who were knighted alongside him but who have maintained more significance in the history of science.[3] An account of his knighthood also introduces themes that were important throughout Bell's lifetime and within the patronage system of London medicine during that period—themes of status and standing within society, looking and dressing the part of a gentleman regardless of one's situation at birth, and being counted a member of a natural philosophical elite.

MEDICINE IN THE NINETEENTH CENTURY

Charles Bell's world, that of nineteenth-century London medicine and surgery, is the subject of this book. If interest is to be found in a world that is enough like our own to be recognizable but that also maintains its strangeness, then nineteenth-century medicine offers a most captivating subject. Descriptions of hospital cases in medical journals contain familiar maladies—kidney stones, tumors, fractures, and cholera—all recognized as such, but treatments bring together that element of familiarity with a certain foreignness. A man enduring a lithotomy, a common surgery for "the stone," was tied with his hands bound to his ankles on the same side, such that he was folded in two. Unanesthetized, he was then bound to the table, and surgery was performed as quickly as possible, both to avoid post-surgical complications (fevers and other signs of what we would recognize as infection) and because the patient was awake for whatever excruciating pain attended the surgery.[4] Some accounts suggest that the amputation of a leg could occur in less than three minutes, and other amputations in less than thirty seconds.[5] Still, mortality rates for a surgery like lithotomy were reckoned roughly one in eight, *if* a surgeon chose his patients wisely and performed well.[6] They might have been as high as one in four for amputations done in a hospital setting (presumably higher on the battlefield).[7] Nonsurgical intervention was also rather violent by modern standards, full of emetics, bleeding, and clysters.

In this world, Bell was attempting to create a science of medicine and to claim for himself standing as a natural philosopher. It was in this world that he earned himself a knighthood. His professional trajectory combines what we might regard as the familiar and the foreign as well. Bell held such apparently recognizable posts as professor and hospital surgeon, and he saw patients privately too; but he also had to canvass for votes for his hospital posts, got paid by the student for his lectures, and was always poor and always in need of patrons. He was in some senses, then, an atypically philosophical and ambitious London medical man, assembling a typical, motley set of jobs in the early nineteenth century in order to make a living as an anatomist and surgeon. It was a world that was on the brink of reform, and the question was only of what sort.

A BIOGRAPHICAL APPROACH TO BELL

This book uses the life of Charles Bell as a sampling device to uncover a strain of what I term "conservative reform" in early nineteenth-century British medical education. For Bell, this vision of conservative reform focused on a pedagogical approach that positioned practical anatomical experience as the key to a specifically British medical tradition. Other conservative reformers might have articulated slightly different visions of medical education, but they shared a belief that British medicine could be improved through incremental changes rather than radical transformation, and most rooted those changes in some form of pedagogical reform.

Perhaps best known among his contemporaries for his priority dispute with the French physiologist François Magendie over the discovery of a law governing the roots of the nerves, Bell was also a medical reformer in a great age of reform, an occasional and reluctant vivisectionist, a theistic popularizer of natural science, Fellow of the Royal Society, surgeon, artist, and (perhaps most importantly) a teacher. Bell's priority dispute with some of his own students and countrymen, as well as with Magendie, makes Bell a figure whose work, reform efforts, and reputation shaped British medical culture at that time, and whose work also helps us to understand that culture.

While Charles Bell was clearly a significant man in his own time, he was neither especially famous beyond his own circles nor venerated after his death. But he is also not an "everyman" of science. Bell built a reasonably successful career as an anatomy teacher, surgeon, and aspiring late Enlightenment philosopher living in an age of reform; and by looking at his individual life, we get a sense for both of those periods and what it might have meant to move through them as an aspiring natural philosopher and ambitious medical man. Because he was reasonably successful, his contemporaries sometimes commented about his work or his personal politics, his letters to his brother were published, and his drawings and specimens were collected and preserved, making him a convenient subject for the historian. This book depends on those letters, some collected in the published volume assembled by his wife after his death and others dispersed in special collections holdings at the Wellcome Library in London, University College London, and the National Library of Scotland, as well as on collections of objects and paintings at the museum of the Royal College of Surgeons of Edinburgh, and on student notebooks at the Wellcome Library and in the special collections at the University of Leeds. The letters in particular make him a likeable character, as one gets to know his insecurities, ambitions, and affections.

Bell's work sits at many intersections, and those intersections make visible not only the complicated and flexible politics but also the interrelated intellectual and academic pursuits of men of this period. In this sense, Bell was very much a man shaped by the late Enlightenment culture of Edinburgh from which he came. While he was a practicing surgeon and teacher of anatomy, he was also an artist and sought a place among that community. He spent time with philosophers and politicians in Edinburgh (such as Dugald Stewart, John Playfair, Sydney Smith, and Francis Jeffrey) and scientists in London (Joseph Banks and Humphry Davy), and submitted his anatomical work to lawyers and philosophers for their opinions, having tried sometimes to engage with their subjects of inquiry as well as his own.

Born in Duone, in Perthshire, Scotland, in 1774 and educated in Edinburgh during that city's heyday of medical and surgical education, Bell received instruction both in the university and by apprenticeship in a private anatomy school run by his older brother John. He subsequently formed part of an exodus of Scottish medical men heading for London in search of teaching opportunities. These Scots hoped to take advantage of London's less unified, more diffuse institutional politics and its multitude of hospital positions and private schools. Never developing a particularly strong private practice, Bell was to hold positions in several of London's medical institutions.

One need not take a psychological approach to see that Bell's career was shaped by his family. Bell's father died when Bell was young, in 1779, but his position as a clergyman in the Scottish Episcopal Church might be seen as related to Bell's fundamental trust in design in nature, an undogmatic but essential element of Bell's philosophy. In a less speculative way, one can assert that in the absence of his father, Bell was especially close to his mother, whom he credited for his skill and training in drawing. Although Bell attended the High School of Edinburgh, he wrote next to a remark in a biography about his education there, "Nonsense! I received no education but from my mother, neither reading, writing, cyphering, nor anything else."[8] The family was poor, by Bell's account, but well connected, and his mother well educated for a woman of her time. We do not know terribly much about those early years (not the people he mingled with in the Edinburgh medical world, nor the classes he might have sat in on), as most of the details of his life are available to us from letters he sent as an adult. The family member who is surprisingly absent from Bell's letters and accounts after he went to London in 1804 is the one who taught Bell his trade—his brother, the surgeon-anatomist John Bell.

One gets the sense that John, who was twelve years older than Charles, was a difficult man. In September 1792, Charles was bound to John through a deed of indenture, apprenticed to him to learn the callings of surgery and pharmacy. In 1790 John had built an anatomy school adjoining Surgeons' Hall to rival the powerful University of Edinburgh medical faculty and, particularly, the anatomist Alexander Monro Secundus. John Bell's was a successful school, one that found its niche by taking advantage of the weaknesses of the Monro dynasty. Bell advertised his class as having to do with the application of anatomy to surgery, something that extramural teachers suggested could not be done properly by Alexander Monro Secundus because he was not himself a surgeon.[9] Surgical anatomy was also one of Charles's strengths, one that he clearly took from John. But John's school was relatively short-lived, and for all that Bell learned from his brother, his brother's reputation was also largely responsible for Charles Bell's exile to London.

In 1799, John Bell closed the school that he had founded after he was banned from the Royal Infirmary in Edinburgh, the one venue in Edinburgh for a student to witness practice in a hospital-like setting. While some have attributed his exclusion from the infirmary to John Bell's success at rivaling the university,[10] it was also clearly the result of Bell's cantankerous personality and long and bitter arguments with powerful university-based medical men, particularly James Gregory. Gregory was the Professor of the Practice of Medicine at the University of Edinburgh, and whatever the original source of their feud, he developed a strong dislike for John Bell. The two engaged in a bitter and polemical battle, sometimes using pseudonyms, in which Gregory accused Bell of plagiarism, and later, after a more direct response, Bell wrote *Letters on Professional Character and Manners: On the Education of a Surgeon, and the Duties and Qualifications of a Physician: Addressed to James Gregory, M.D.*, leaving nothing about the identity of his target to the imagination.[11] The book is over six hundred pages long and bespeaks a long-running and consuming feud.

Edinburgh was a small and highly politicized city and the College of Surgeons in Edinburgh formed an integral part of the town council.[12] Thus, politics and professional life clearly mixed. When John Bell was excluded from the infirmary, it became abundantly clear that he was on the wrong side of those university factions within medical politics that possessed power within the town. He explained to the council "how inseparably connected his system of teaching was with the best interests of the patients, as well as with the improvement of surgery,"[13] and thus, committed to his principles and true to his word, when deprived of his privileges there he shut his school's doors.

And so Charles, whose political fate was sealed by his brother's enmities, went to London. One might imagine that the conflicts of his brother, which ultimately drove Charles out of town, shaped Bell's aversion to such quarrels later in life, when he refused to engage in a dispute begun by Herbert Mayo, his former student. Charles Bell said of his approach to conflict: "I left controversy to my younger friends . . . in spending the greater part of my life in the duty of teaching, I had educated many to the profession who knew the correctness of my statements, the mode and succession in which my ideas had developed themselves, and who were as willing as capable of defending me against illiberal attacks."[14] But Charles took more from Edinburgh than his aversion to conflict and awareness of local politics (though not his casebooks and sketches, with which John refused to part).[15] He began his publishing career there. Charles and John published a four volume *Anatomy of the Human Body* together, the first volume, by John alone, appearing in 1793, the later volumes and engravings being done solely by Charles.

LATE ENLIGHTENMENT EDINBURGH

Charles's early interests were clearly shaped by the Edinburgh milieu and the university faculty. His Edinburgh was that of the late Enlightenment, and the eminent philosophers who surrounded him were Dugald Stewart, Adam Smith, Francis Jeffrey, and John Playfair. Bell had studied at Edinburgh at a time when moral philosophy was at the core of Edinburgh's Enlightenment. Sydney Smith, in an 1872 profile of Bell's "Letters and Discoveries" for the *Edinburgh Review*, wrote of Charles and his brother George

Joseph's early education that "the three men who more than any other determined their future course were John Millar, John Playfair, and Dugald Stewart."[16] Furthermore, Bell referred occasionally and admiringly to Stewart in his letters to George even after he had moved to London.[17] While that background helped to shape Bell's focus on the mind, it is hard to see any sort of direct "influence" argument pertaining to Bell, at any point in his life, even if one were inclined to try to make one (and I am not). Instead, Bell's early life among Edinburgh's philosophical elite framed his understanding of the place he did or should inhabit among philosophers. Unlike many London medical men, Bell imagined himself a part of a philosophical, rather than a practical medical, community from the outset.

It was no accident that Bell set out to make a discovery regarding the brain and, through that path, ended up working on the nerves. Bell inhabited a social and intellectual world in which questions about the ways in which the brain worked were of central concern.[18] Bell's Edinburgh would be home to phrenology, a subject that Bell himself did not take seriously, but that also betrayed a preoccupation with the brain. But concern with the brain predated the rise of phrenology and extended beyond circles of anatomists. David Hume, Thomas Reid, and Dugald Stewart built philosophies of mind, each helping to shape the character of natural philosophy in Edinburgh. They debated questions of perception and of the relationships between the mind and the body and between the mind and its physical organ, the brain, questions that also took on great significance for Bell, whose anatomy and pedagogy both addressed such topics.[19] Hume and Adam Smith were close friends with another Scottish medical man, one whose career predated Charles Bell's, but whose role in shaping Edinburgh medicine and its focus on the brain was central: William Cullen.

Cullen, who had taught William Hunter for three years, began lecturing on medicine in Edinburgh in 1757 and remained in that city until his death in 1790. He was a pedagogue, a systematist, and a skeptic, and he reformed a medical curriculum that had been based on the work of Herman Boerhaave.[20] Bell's Edinburgh was the one shaped by Cullen, who had regarded the nervous system as central to physiology and pathology and who, according to Christopher Lawrence, was in dialogue with Hume when he proposed that mental impressions be divided into sensations and ideas. Theories of sensations and impressions on the mind came to preoccupy much of Bell's work and his teaching philosophy.

Like the London surgeon William Cheselden before him, Charles Bell attempted to resolve John Locke's question about the relations among seeing, feeling, and knowing.[21] And Bell's Bridgewater Treatise on the hand argued that sensations—experiences with objects—were the foundation of learning, followed by language. It is hard not to see in such a pedagogical philosophy traces of the so-called Common Sense Philosophy of Edinburgh and of Dugald Stewart, who wrote, "how far soever we may carry our simplifications, we must ultimately make the appeal to facts for which we have the evidence of our senses."[22]

While Bell's natural philosophy was in some senses fundamentally Scottish, his medical politics and immediate medical community were distinctly those of a Londoner. Edinburgh medical men, after all, had driven him out and forced him to London. Moreover, Edinburgh had always been a city in Britain to which continentals flocked, bringing with them their French-style medicine. When Bell attempted to shape or define British medicine, therefore, he was talking about London medicine and anatomy, for it was in London that he built his reformist pedagogical platform.

CHARLES BELL'S MEDICAL LONDON

While Charles Bell's Edinburgh was nearing the end of a long philosophical renaissance, medical London was something of a practical mess, being the capital of a country whose medical system was at best disorganized and at worst corrupt. Until the late 1820s in Britain, according to the *Edinburgh Review* in 1845, "there were nineteen different modes of obtaining a licence to practise medicine, nineteen different methods or forms of education to prepare for that profession, and fourteen varieties of privileges attached to it." That meant that medical licensing was handled differently everywhere, as a nineteenth-century commentator observed: "in this anarchical institution, the highest degree, that of Doctor, was conferred in one place by the universities, in another by the colleges, in a third by a single person, such as the Archbishop of Canterbury, whose favour was worth ten years of study."[23]

In London, the Worshipful Company of Apothecaries, the Royal College of Physicians, and the Royal College of Surgeons each credentialed practitioners in ways that filled their own coffers without assessing competency to practice.[24] While most medical men in London acted as general practitioners, doing a bit of surgery, a bit of medicine, and a bit of drug preparation, licensing and training alike divided the medical world into three distinct groups with strong suggestions of place in a hierarchy attached, and credentialing for each involved a different curriculum (and a separate set of fees!). Apothecaries (druggists) were at the very bottom of that ladder, with surgeons in the middle, and medical doctors at the top. These sorts of divisions did not actually fit patterns of general practice, in which surgeons, medical doctors, and apothecaries all tended to diagnose, dispense medicines, and perform operations when necessary. They were particularly problematic for apothecaries, who, to abide by licensing laws, were to choose between offering medical advice and dispensing drugs (something that rarely happened in practice). Licensing therefore became an easy target of criticism for reformers of all stripes, since it also tended to dictate the kinds of training students sought.

Given such disarray, most historians do not see Britain as the center of the nineteenth-century medical world. For them it is late nineteenth-century developments like the use of systematic autopsy and accumulated medical data, anesthetics, and the introduction of laboratory procedures, some of which were imported to Britain from the Continent, that signaled the development of scientific medicine in Britain. [25] But Bell's London was a slowly evolving one, full of activity, particularly in the schools.

Famed teachers and practitioners like John Abernethy, Astley Cooper, Benjamin Brodie, Richard Bright, and Thomas Addison taught practical medicine in the charitable hospitals of London, where the progression of diseases in the sick poor was carefully studied and marked. Thriving private schools educated students on a variety of medical subjects. Despite the lack of a central university in the nation's capital, it becomes clear, when examining medical and surgical education in London at the beginning of the nineteenth century, that, as Roy Porter put it, Georgian Britain was not the "wasteland sometimes supposed."[26] In London, private schools flourished, as did hospital training,[27] but perhaps the greatest innovations in London medicine during the early nineteenth century came through pedagogical reform.

The Great Reform Act of 1832—an act of Parliament that was proposed by the Whigs and extended the franchise, reshaping electoral politics in Britain[28]—was passed as Bell's career hit maturity, a year after he was knighted and a year before his Bridgewater Treatise on the hand was published, and has caused historians to define the surrounding decades (as early as 1780 and late as 1860) as the "Great Age of Reform."[29] In the Age of Reform, all institutions seemed subject to the reform impulse, and medicine was no exception. A variety of medical reform movements existed in the early nineteenth century alongside reform movements of other sorts, including those that were political in nature. Medical reform generally incorporated licensing, institutional, pedagogical, and sometimes epistemic reforms, all of which were intertwined.

It is important to declare from the outset that I will address these reforms and the social interactions that supported them as a form of "politics," but that whether I am talking about conservative or radical reforms, about personal and professional networks, or about personal or institutional politics, I am not talking about ideological, party-based politics. The stories of Whigs and Tories do not correspond to those of medical men, whose politics were situational and based on professional relationships and ambitions, so the words "conservative" and "radical," or the word "politics" should not be anachronistically read through the veil of twenty-first or even nineteenth-century ideological political parties. In some senses, that is the very point of this book, with its biographical focus to help reveal that politics were built on personal networks and that they included espousing seemingly contradictory positions as the situation dictated.

Some Londoners sought radical medical reform. Led by Thomas Wakley, a surgeon, the founder of the *Lancet*, and later a Member of Parliament, these men, termed "radical reformers" by historians, wanted to import continental life sciences like physiology and morphology, and philosophies like materialism, with their threatening associated atheism. They also wanted to overthrow and reconstruct institutions of power in London, like the Royal Colleges of Physicians and Surgeons, leveling hierarchies as the French had done in the period following the Revolution and establishing a meritocracy that might better serve lowly general practitioners.[30] Their counterparts, a group that I call "conservative reformers" because they employed a rhetoric of tradition in their calls for reform, saw some room for improvement in the British medical world, but argued for gradual change in educational requirements and licensing laws that would promote more comprehensive schooling in practical medical sciences, and for small modifications to the governing professional institutions already in place, rather than revolutionary overhaul of medical institutions.

Although true medical conservatives, that is, nonreformers who saw no need to implement any changes, did exist—often at places like Oxford and Cambridge, where medicine remained a theoretical and Latinate discipline rather than a practical one—they were relatively rare in London. But the much more sizeable group of conservative reformers sought moderate reform, reform that *conserved* what they took to be, or construed as, British tradition while making hierarchy more sensible, training more practical, and pedagogy more systematic. The systematization of medical training undertaken by London medical men like Bell was a genuinely reformist and innovative move, one with social, structural, and professional ramifications.

To understand the politics of medical reform, one must look at individuals. Men like Charles Bell were part of natural philosophical and university communities, of local networks of medical men, surgeons, and scholars, of networks of patronage, and of political networks. Their politics and priorities were multifaceted and flexible, in part because they had to be—there were no "career paths" on which these men could travel, so they cultivated connections and seized opportunities (both academic and financial) where those opportunities appeared. Sometimes political sensibilities, as they are acted out, reside so much in the individual, with so little coherence among individuals, that political ideologies can be read only as the landscapes through which individuals moved and not as proper to individuals themselves. In Britain, the early nineteenth century is one such time, and Charles Bell demonstrates why. While "politics" can be taken to mean ideological politics—political positions—this biographical approach allows one to take seriously the idea that ideological political positions are also constituted by and constructed alongside social interactions. Conservative reform, which disappears from the historical landscape if one is looking for an ideologically driven political group, is made visible when one focuses on the individual.

"Conservative reform" is a concept with wider purchase than this specific case. It aptly describes Humphry Davy's stance toward institutional reform[31] and is similar in nature to Michael Gordin's description of the innovative but nonradical Dmitrii Mendeleev, who also tried to "preserve traditions essential to Russian stability" in the face of revolution.[32] In periods of revolution, antirevolutionaries seem often to call for conservative reform, for holding something steady, returning to something they know to be safe, even while promising change.

It will be noted that the French often emerge as a sort of secondary subject in this book. When they are mentioned, they are always mentioned as Londoners saw them—and, as invoked by Londoners during the early nineteenth century, they were either heroes or villains. It is doubtful, of course, that they were actually either. But this is a book about London medicine, and others have written about the actual state of French medicine during this period.[33]

LONDON CLASSROOMS AS A REFORMER'S STAGE

Nineteenth-century medical science in Britain was fundamentally about teaching. The classroom (a term I am using in an encompassing way, so as to include not only lecture rooms but also wards and museums where teaching was conducted) was the space that brought together all of Bell's endeavors, and it unified reform efforts as well as scientific pursuits generally in London medicine. The classroom was the space in which anatomists' careers were made, their audiences built, their research conducted, their work published, and, centrally, their pedagogical practices developed. Overlooking the classroom and pedagogy in favor of the separate spaces that characterize modern medicine—the lab, the clinic, and the journal—is to overlook much of what was important to early nineteenth-century Britons. This book restores the classroom and its associated activities to their rightful place at the center, tracing the emergence of a reformist pedagogical program through Bell's classroom-based research and teaching endeavors.

Chapters move roughly chronologically through Charles Bell's life in London. The first chapter, "Politics and Patronage: Building a Career in London's Medical Classrooms," addresses Bell's early years in London, including his attempts to develop his anatomical research and make a name for himself as a natural philosopher, to cultivate patronage and build a private school there. While creating a system of patronage and support through the social and intellectual network of Scottish men in London, Bell attempted two broad systematizing projects with which he would build his reputation: one a system of the nerves and the other a system of education. The second, "Pedagogy Inside and Outside the Medical Classroom: Training the Hand and Eye to Know," moves inside Bell's classroom, examining the ways that he conceived of pedagogy and the ways in which he incorporated various kinds of displays and objects into his teaching. Charles Bell's medical pedagogy and his medical science were both developed within the classroom, through the creation of visual objects meant both to constitute the science—describing anatomical structures and their logic through imagery and its accompanying text—and also to teach it. Chapter Three, "From the Anatomy Theater to the Political Theater: Journals and the Making of 'British Medicine' in Early Nineteenth-Century London," addresses politics, pedagogy, and the emergence of weekly medical journals. Politicized journals attempted to bring together a wide-ranging community of readers and created a space within which the ideologies behind medical education were crafted to have specific political implications. Chapter Four, "London's New Classrooms: London University and the Middlesex Hospital School," follows Bell in the 1820s and 1830s, as the institutional settings of London medicine changed and new universities and hospital schools were being founded. And finally, the fifth chapter, "Defining a Discovery: Changes in

British Medical Culture and the Priority Dispute over the Discovery of the Roots of Motor and Sensory Nerves," traces the priority dispute between Charles Bell and François Magendie, a dispute that extended over the entire second half of Bell's life in London, and that made Charles Bell famous in his own time. The dispute offers insight into the changing contours of the medical world during Bell's lifetime; Bell's own medical science, born in Enlightenment Edinburgh, became a part of the dispute itself, and while Bell did not lose the dispute per se, the style of medical science Bell hoped to defend was no longer viable by its end, at the close of Bell's life.

On March 2, 1818, Charles Bell wrote to his brother George, "If I am to be anything, it is from connection with Natural Philosophy by Anatomy." He was among the last of a generation of natural philosophers who were superseded by a generation of laboratory "men of science." Bell built his philosophy in Edinburgh, his pedagogy in London, and his political network across geographical, class-based, and political lines to become a knight, a founder of educational institutions, Fellow of the Royal Society of London, practicing surgeon-anatomist, and the British discoverer of the separate roots of motor and sensory nerves. This is his story, and with it, the story of medical science and medical reform in London's classrooms at a time when modern medicine, with its practical universities with set curricula, staffed by medical professionals, was being born.

CHAPTER ONE

Politics and Patronage: Building a Career in London's Medical Classrooms

I have got your scolding letter to-day. You tell me to cultivate men; I wish you had said, to be industrious and cultivate a proud spirit of independence.

CHARLES BELL TO GEORGE BELL, May 1809.[1]

On November 28, 1804, Charles Bell arrived in London from Edinburgh, alone. Just two days later, he wrote to his older brother George, "For these two days I have not been idle. I have called on Baillie, Lynn, Thomas, Wilson, Abernethy, Cooper, Gartshore. I believe I shall find every attention from the medical people here."[2] When Charles Bell embarked on a career in London, he left his brother, his dearest friend and confidante, behind. That letter was the first in an almost-daily correspondence between Charles and George that lasted throughout Charles's many decades in London; George, a lawyer, also helped Charles financially. Their letters are candid and offer insight into the ordinary attempts to fashion a career[3] by a man who would become, but was not yet, famous. Those letters demonstrate what Charles understood himself to be doing, as conveyed in a sort of "real-time" narrative to his brother, providing evidence of how networks and patronage, as well as politics, were built during the period. His initial letter to George describes a lifelong pursuit of Bell's, one that was advised by George in the epigraph to this chapter—to cultivate men as patrons. With his entry into the London medical scene, Charles Bell began his initial attempts to build networks of patronage and develop a career.

PATRONAGE IN LONDON

While some have pegged Bell as a conservative Whig,[4] his politics are actually much harder to pinpoint, and, as this book will demonstrate, are not revealed by looking at ideological politics. Like many of his contemporaries, he was what I call a conservative reformer, whose politics were situational and not ideological. He was not without beliefs, associations, or ideals, but, as this chapter argues, his world was one that rewarded adaptability. Bell's friends were Edinburgh Whigs, to be sure, but despite these personal connections and his own political leanings, Bell watched the London political scene as if from a distance, commenting, for example, on the style but not the content of Charles James Fox's address against Lord Melville.[5] In one passage from 1810, Bell seems to sum up the necessity of making oneself a well-polished gentlemen without ideological politics. He had been asked to have wine at the Beef-steak Club with the Duke of Sussex, the Duke of Clarence, and the Duke of Norfolk. Bell describes the evening, mentioning his attire, saying "[f]inding myself thus unexpectedly situated, with my pale face and black coat, and knowing the lean of these princes, I gave [toasted] the Chancellor of Oxford, and passed for a staunch supporter of the Grenville interest there, which was well received."[6] In other words, Bell gave the dukes to believe that he shared their politics. The Duke of Clarence signaled his approval, according to Bell, "saying (observe the familiarity), 'Mr. Bell, I hold a good toast to be better than a good song, and I drink this with particular pleasure.'"[7] Bell ends his account of the evening by commenting on the hard work involved in maintaining patronage: "The manners of the princes are certainly admirable, but it must be easier to be a polite prince than a polished gentleman.[8] Bell, knowing the "lean" of the princes, had passed for what they wanted—"a staunch supporter of the Grenville interest," but unlike polite princes, who simply had to demonstrate good manners, the polished gentleman had to work to dress the part and to share in the princely good manners, but also to know his audience and please them. Gentlemanliness was a necessary professional act that required the appropriate dress, residence, dining establishments, and friends, and a politics to suit all occasions.

Such political detachment, agreeableness, and flexibility were useful to Bell, who spent his time in London fashioning himself as a character suited to the multiple audiences necessary to make a career. Part gentleman natural philosopher, part artist, and part practicing surgeon, he modeled himself in such a way as to attract powerful supporters among the scientific and social elite, to fill his classrooms, and to enhance his social standing—the necessary pieces that could be slotted together to make a career in the capricious world of early nineteenth-century London medicine.

Medical patronage in the eighteenth century was shaped by relationships between practitioners and their clients—wealthy gentleman with idiosyncratic theories and no medical training.[9] By the beginning of the nineteenth century, patronage and networks were still important, as there was no defined avenue to professional success for a London medical man, but the nature of that patronage had changed. Bell sought support first from established medical practitioners rather than from wealthy patients, and then later from learned gentlemen. While there was still no career path for medical men, no set of professional positions that guaranteed sufficient income, the nature of patronage and of medical work had changed such that Charles Bell could imagine that, with the right friends, he would be able to build and sustain a career teaching in private schools and the wards of hospitals, publishing, and illustrating the medical texts of others.[10] Such a career required the constant and long-term maintenance of a variety of networks. When Bell arrived in London, he immediately began to build these sorts of connections.

One such system of friends and source of support, and one in which Bell tended to feel at home, was made up of fellow Scotsmen. The Scottish community in London provided Bell with an important base, and it was from that community that Bell received most of his backing from those whose professional lives were outside medicine. Homesickness is readily apparent in Bell's letters to his brother, to whom, shortly after arriving in London, he wrote, "I still, in all my wanderings of imagination, keep my eye fixed on home; and the hope is always prevailing that I shall be able to collect such information as may entitle me to return to you with comfort to us both."[11] In addition to missing his family, Bell's thick Scottish accent often gave him trouble in English company, so he seems to have sought and found sympathy and ease in the company of other Scots. In January 1805, he wrote to George about dining at the Edinburgh Club with, among others, the writer and editor for the *Edinburgh Review*, Sydney Smith. Even at the Edinburgh Club, despite evident support for his work (he reported Smith "delighted" by it), he found himself subject to anglicizing. His report to his brother: "He [Smith] has received much pleasure, and has been much entertained. But, what do you think?—he proposes that I should put it [*Essays on the Anatomy of Expression in Painting* (1806)] into the hands of a literary hack to brush away the Scotticisms!! He further tells me that Mrs. Smith has read it, and marked several Scotticisms. Be this as it may, he is to return it, I believe, with a letter . . . I am a little mortified about these Scotticisms."[12] Even the wife of a friend and fellow Scotsman took him to task for his unpolished writing, the reform of which would be just one element of acting the part of a successful surgeon in early nineteenth-century London. In July 1805, he wrote to George, "I dine to-day with Longman,—all Scotch,—Horner, Brougham, Allan, S. Smith, Abernethy. Nobody will interfere with my language!"[13] But, by 1827, the *Lancet* was still able to mock his pronunciation of the word "spicimin."[14] Bell remained always a Scotsman in London.

In addition to building that system of Scottish connections, Bell also immediately and diligently began building a network of medical practitioners—well-connected physicians and surgeons who might serve as professional patrons, helping him to secure a professorship or hospital position. Bell's first letter home, already cited, mentions calling on "Baillie, Lynn, Thomas, Wilson, Abernethy, Cooper, Gartshore."[15] Their names are largely familiar to historians of medicine—Matthew Baillie was the nephew of William and John Hunter and is often recognized for his contributions to pathology;[16] William Lynn was a surgeon to the Westminster Hospital;[17] James Wilson was the proprietor of the Great Windmill Street School of Anatomy;[18] John Abernethy was, by that time, such a famous lecturer that the governors of St. Bartholomew's Hospital had built him a theatre;[19] and Astley Cooper was an English surgeon-anatomist who practiced at Guy's Hospital and had recently received the Royal Society's Copley Medal.[20] These were the leaders of the surgical community in London. Bell pursued relationships with such men as an occupational necessity, and he did so in a variety of ways. His early letters are brimming with references to instances in which he arranged to dine with those whose patronage he sought.[21] Cooper, Baillie, Abernethy, and Lynn, together with Henry Cline (a surgeon at St. Thomas's and Fellow of the Royal Society[22]) became Bell's regular dinner companions in those first London years. Such meals were social, rather than professional, occasions. In one 1806 letter to his brother, for instance, Bell wrote, "I went yesterday to dine with Abernethy, and in the evening to Vauxhall . . . The night was pleasant, the scene brilliant, and my party, though not exceedingly interesting, agreeable. I had generally Mrs. Abernethy and two little ones under my care, and was quite *en famille*."[23] While ostensibly social affairs, the professional significance of such outings was never far off.

Bell also built his network of benefactors among the medical community through interactions of a more professional nature. To show off his natural talent, interest, or expertise, Bell offered his services or demonstrated eagerness to learn in a variety of ways. He often assisted or watched the work of established surgeons. In an 1805 letter, for example, he wrote: "Today I took a long walk into the City . . . I was seeking Sir Charles Blicke, one of the surgeons in St. Bartholomew's; missed him, left my card . . . I meant to say, 'Sir Charles, I am come to study from men, having made the most I can of books, and I am going to take the freedom of attending all your operations.'"[24] Bell also assisted William Lynn with surgical operations, continuing to follow patients with Lynn in the hospital until their death or release.[25] Sometimes he described his role as a mere "spectator" at such operations as agonizing and "torture," and bemoaned the state in which surgery was practiced.[26] But he nonetheless made a point of using opportunities to witness the work of more practiced and better-connected surgeons in order to gain favor and friendship with men he hoped to make into colleagues.

In addition, Bell used his exceptional artistic skill as a way of introducing and ingratiating himself. He made drawings for Astley Cooper and also for a book on the arteries by James Macartney, surgeon and lecturer on comparative anatomy and physiology at St. Bartholomew's Hospital. Of this work for others he wrote: "I wish to oblige people, and to have it said that I *am* willing to give them the assistance of my labours."[27] Lending his hand and his artistry to their work, Bell hoped, would allow him to establish a reputation. He had his eye as much on the position of anatomy professor at the Royal Academy as on those of London's hospital surgeons.

Seeking champions and advocates was not incumbent only upon the newly arrived surgeon—maintaining a network of patrons was a career-long and career-shaping pursuit. Whenever a position Bell wanted opened up at a hospital, college, or academy, he made the rounds to those he knew, canvassing for votes, asking patrons to call on their friends on his behalf, and he did so throughout his career.[28] Benefactors from outside the profession helped finance books or museums, and in Bell's case, those relationships resulted in strong ties, such as publishing ventures between Bell and Henry Brougham, a politically connected Edinburgh Whig who had interests in education.[29] Sometimes the pursuit of support from successful medical men and wealthy lay people happened conjointly, as in a dinner Bell described: "Abernethy was with me to-day, and we talked a great deal pretty openly; he renewed his protestations of regard, &c. . . . I dine with him on Thursday with some foolish lord."[30] Here, professional support offered the possibility of further patronage from "some foolish lord."

But establishing a reputation and a structure of support—calling on men, dining out with them, offering one's services and assistance generously to professional colleagues—was costly. Doing so meant dressing to suit one's aspirations (rather than one's circumstances), living in the right kind of house at the right sort of address, and eating at the right kinds of establishments. In 1805, Bell moved to Fludyer Street in Westminster, for example, because the street was, he said, "respectable and genteel, and in my apprehension, just such a one as I ought to be in."[31] And to trace Bell's discussions of his coats is to understand the significance of such trappings of success. In his first months in London, in a letter complaining of poverty, Charles wrote to George, "I have given myself a new hat, coat, and waistcoat."[32] That might be expected of a young man trying to prove himself, but in 1822, almost two decades later, after having published some of his most important works and shortly before he was named Professor of Anatomy and Surgery of the Royal College of Surgeons, he wrote to George: "Last week I went to Sir Humphrey [*sic*] Davy's meeting, and there I found my paper had done me as much good as if I had bought a new blue coat, and figured French black silk waistcoat."[33] Fashion and social standing, as well as professional position, surgical skills, and patients, remained important elements of a surgeon's life throughout his career in Regency London.

Living like a gentleman, though, left Bell barely able to keep up with the bills. In 1807, he wrote to George, "I have another hundred in my coffers, but with a breath it will melt again. I have taxes, rent, and accounts to pay that will require it and more."[34] Bell wrote often to his brother to ask for money to support the activities that he meant to establish his standing and secure him a place in the London scene. In one such instance, for example, he pleaded "I really cannot do here without more money . . . I see incalculable occasions of expense in establishing myself, but no other obstructions."[35] This was the cost of doing medical business in early nineteenth-century London.

ESTABLISHING A CLASSROOM

While Bell was developing a network of patrons, he was also busy establishing himself as a teacher and cultivating an audience of paying students. Surgeons like Bell cobbled together income from a variety of sources, and they often supplemented medical practice with teaching, an ordinary and reliable additional means of earning a living. Independent lecture courses sprang up near to and in conjunction with charitable hospitals whose wards provided additional experience to students (and an additional source of income for those hospital-affiliated teachers whose students paid to "walk the wards" with them).[36] As Bell himself detailed in a letter, St. Thomas's and Guy's hospitals, located near each other on the south bank of London, were thriving and had many lecturers and students, but where John and William Hunter had previously reigned—the former at Leicester Square and the latter at Great Windmill Street—on the north side of the Thames, there was only Thomas Wilson, successor to Matthew Baillie at the Great Windmill Street School itself. William Hunter's museum had gone to Glasgow, and William Cruikshank (also of the Great Windmill Street School) had sold his collection to the Emperor of Russia.[37] It was thus that Charles Bell, with the help of George, who visited eight months after Charles arrived in London, decided to buy a property on Leicester Square at which he might set up a museum and classroom.

The decision to buy the house on Leicester Square was one that came at a watershed moment for Bell, a moment in which he decided to persevere in London as a private anatomy teacher, despite his despair at not yet having found success in a regular hospital post. Bell later wrote of that decision: "One night I resolved to return to Edinburgh. I went to the Opera, to leave the last pleasant impression of London. I could dwell upon my feelings of that night, but few could sympathize with them—and next morning I resolved to remain in London." He accordingly secured a large house in Leicester Square, despite its poor condition. "When I went with my surveyor to examine it, I was some what appalled by his account; he was a great John Bull rough fellow. Leaning out of the window, and observing the walls out of their perpendicular, he said in a coarse, familiar manner, 'Sir, you had better have nine bastard children than this house over your head.'"[38]

The house at Leicester Square, despite its "ruinous" condition, was such a financial burden that Bell, in buying it, reaffirmed his need to give "[him]-self up exclusively to the teaching of anatomy."[39] Teaching was the central conceit of the rest of his life, and its philosophy can be identified in and unites almost all his published work. And so Bell set about creating his lecture theatre and associated museum at Leicester Square and filling them with students.

George sent down all of Charles's books from Edinburgh at that point, as well as packages filled with preparations (specimens and models) that Charles had made. George also handcopied Charles's casebooks and sketches, which Charles had done when he was an apprentice to his brother John, who now refused to part with them. Charles took on two house pupils as apprentices, and they helped set up cases and varnish newly arrived specimens. The museum was filled with fairly standard sorts of examples of human anatomy, as well as with oddities and animal specimens. Charles wrote to George about sending a hippopotamus's head[40] and about having acquired "the skull of a Roman, with an obolus in his mouth, and a very curious diseased bone belonging to the same skeleton."[41] He also acquired specimens from other surgeons and predicted, "I shall soon be universally known, and my museum will increase rapidly."[42] When George saw the newly built museum for himself in July 1807, he wrote of it to his wife, "Many people come to see his museum, but it is not yet so far advanced as the labour and expense bestowed on it I hoped would have made it. However, it only requires him to go on with it as he has begun; and I am told, that even now, it is one of the most interesting things of the kind in London."[43] But, however much the museum and its contents served as a public attraction, displays were an important part of Bell's anatomy teaching, and the museum was built primarily for Bell's students.[44]

Immediately upon moving into the house at Leicester Square, Bell began recruiting students for classes that began right away. On September 30, 1805, his first day as an independent anatomy instructor in London, Bell wrote to his brother, "The classes are all beginning to-day, and I have only my door painted! That's hard. Don't send anything more by mail till I think and feel what I want."[45] After that point, Bell's lectures, and their success or failure, became frequent topics of his letters. Success in the free medical marketplace in which Bell worked was judged as much by the number of students as by the content of the lectures or the students' comprehension of them. In his first report home, Bell wrote, "I found about forty in the room; my disappointment, I must confess, was extreme. The lecture, I believe, went off well, or the people about me are most affectionate liars . . . As yet only three pupils have entered their names."[46] Students had many choices, and in this instance, while forty had showed up to hear his lecture, only three had signed up for Bell's course and paid for the series. The numbers were to continue to plague him, and Bell relied on the network of medical friends he was cultivating to help direct students to him.

By the time Charles Bell moved to London, teaching classes on anatomy or medicine had become a regular part of building a complex medical career; in fact, that was so much the case that competition had become fierce. Contemporary advertisements suggest that in London, a city with no university or central medical school, there were thirty-eight teachers offering medical classes in London in 1800 and forty-four in 1814.[47] These classes were organized into a sort of accepted "timetable," based on when specific kinds of instructors were free, but also in such a way that students could take multiple courses from multiple instructors at the same time. "Anatomy" and "Dissection" were usually taught in the afternoon and "Principles of Surgery" was taught in the evening.[48] An early advertisement from

Bell's Leicester Square days suggests that he kept to that schedule himself, promoting a surgical class that began after the anatomy class each day and was continued on Tuesday and Thursday evenings. Dissection rooms at Bell's school were open from 8 a.m. to 10 p.m., with demonstrations at 1 p.m. Hoping that deals would attract students, Bell offered discounts and incentives for those taking more than one course with him: first course, five guineas; second course, four guineas; third course three guineas; or as many courses as a student liked for ten guineas up front.[49]

Charles wrote to George concerning his first two lectures, saying of them, "They meet with much approbation, but I am not destined yet to see an end to my anxieties and disappointment. There were only twelve pupils, nay, I should say attendants; the day has produced 10 pounds."[50] Such a report might have been discouraging, but Bell wrote that he was not concerned or discontented: "I have labour, that is occupation. I have a few very kind fellows about me. I have not only their kind wishes but their respect, and the three months will, I hope, pass over very pleasantly, and be attended with an increase of reputation."[51] He had begun to realize, it seems, that his classes, much like his connections to supporters in the medical and surgical communities, needed to be built and developed through reputation, over time.

In addition to teaching courses for anatomists and surgeons, Bell, also an artist, was able to provide classes for painters. Hands-on dissection was, Bell thought, an important part of training artists, and one that was lacking at the Royal Academy. The artists often provided satisfaction and enjoyment to Bell: "On Wednesday evening I gave them a general view of the system of a limb, bone, muscle, tendon, of arteries and veins. This is excellent exercise for me. They stand with open mouth, seeming greatly delighted."[52] They also provided a nice supplemental income, as he noted, saying: "There were originally six came to me; they gave me 10*l*.10*s*, and pay all expenses, but they requested me to admit more, and I have twelve at two guineas a-piece . . . My surgical pupils have brought me 82*l*.; my painters will give me 25*l*. A very little further success, and two classes in the year, will be something considerable."[53] By the spring of 1806, in his second term of teaching, Charles Bell was beginning to find success as a teacher. To his brother, his best source of financial support, comfort, and friendship, Charles wrote: "Nothing, my dear George, could give you more satisfaction than to see me—to-day for example—in the midst of my students. Below there were anatomical and surgical pupils at work, and I in the midst of them. In my great room seven painters, with each their little table, drawing from the skulls and skeletons. Then came my public lecture, and then idleness, heightened by the labours of the week *well* performed."[54]

Things continued nicely for Bell. In the summer of 1807, he arranged to take on fifteen-year-old John Shaw, the brother of George's wife Barbara, as an apprentice.[55] And in the fall, Bell's second autumn teaching at Leicester Square, Charles wrote to George, "I made John Shaw write to you to-day to tell you that at my introductory lecture all went well. I gave a very so-so lecture to a good class—that is to say, the little theatre which I have made to hold thirty-five, was overflowing. I gave out ten tickets before lecture, and I never in any former year gave one."[56] Success was starting to pay off. Where Bell had made just over one hundred pounds in his second term teaching, he more than doubled that in the fall of 1807. "On looking to my book of fees for your information I find I have received 283*l*. 14*s*. in the last two months,"[57] he wrote to George. Still, building a class depended on reputation and standing, and while his class on surgery that fall was full, the anatomy class was a source of frustration: "I have been disappointed on all hands. Everybody had pupils for me, but no one turned up! I am thus by character slowly getting a class, but it is uphill work."[58] Bell clearly depended on referrals from other surgeons, doctors, and anatomists whom he had met since arriving in London, relying on his friends to spread the word to potential students who might fill his classroom, though he had less time by then to devote to his network of patrons, and sometimes those patrons disappointed him.

A career of teaching kept Bell busy. He supplemented his work in the classroom with what patients he could attract. One daily to-do list reflected well the lifestyle of a surgeon trying to make a living in London: "After breakfast, to see a patient at Brompton Row, a mile beyond the end of Piccadilly: come home and study for my lecture: after lecture, visit three patients: dine near six: after dinner, prepare for my evening lecture: and in the evening I'll probably have to go to Brompton again."[59] Another account finds Bell similarly occupied, though in this instance with social as well as professional business—an example of the ongoing nature of networking, even amidst a crowded schedule, when social ventures had to be fitted in: "I have been very busy to-day; took twenty-four guineas in the morning, gave a long lecture, dined with a party at Brompton at Col. Baillie's, and returned home at eight o'clock to my evening lecture in good style to a very crowded class—a respectable number at least. Now I am going to bed thoroughly fatigued."[60] But that very letter describing fatigue also describes an additional facet of Bell's professional life in those early years in London: writing.

In addition to maintaining a grueling teaching schedule and seeing patients, Bell was working on *Essays on the Anatomy of Expression in Painting* (1806), *A System of Operative Surgery* (1807 and 1809), *Idea of a New Anatomy of the Brain* (1811), and *Letters Concerning the Diseases of the Urethra* (1811),[61] all of which were published during Bell's first years in London, before he bought the Great Windmill Street School of Anatomy in 1812. In writing to George, Charles ends the letter that began with a fatiguing day by turning to his own work. He recognized with excitement that his work as an anatomist, conducted among the classrooms, was developing into something characterized by greatness. "My new Anatomy of the Brain," he wrote "is a thing that occupies my head almost entirely. I hinted to you formerly that I was burning, or on the eve of a grand discovery . . . as interesting as the circulation or the doctrine of absorption. But I must still have time. Now is the end of a week, and I will be at it again."[62]

THE CENTRALITY OF TEACHING

In addition to teaching classes for the untrained, Bell took on the close supervision of house pupils who also helped Bell with his dissections and course preparations. House pupils, somewhat akin to apprentices, lived with their teacher and learned from him. For that privilege, they paid Bell, contributing to his income, but those house pupils also bound together a tight network within Bell's home, amidst a more dispersed London network. More than one of Bell's house pupils became an anatomy teacher himself.[63] They also served more mundane functions. Of John Shaw, a relative of George Bell's wife, Bell wrote: "He has been a great want. I don't know where to find my lectures: he had a sharp eye after my bits of paper."[64] More importantly, they helped to provide a community within Bell's home, creating a sort of workshop atmosphere, with Bell leading the group, his house pupils supporting and contributing to Bell's work, and the students learning from Bell, the master. Bell wrote, "my house-pupils have been my means of connection with students at large . . . Hitherto my reception among all my pupils is most gratifying, and this I attribute to the influence of house-pupils, as I never experienced it before I had them . . . Even this winter their merry voices and friendly countenances have been my greatest comfort."[65]

In addition to the camaraderie they fostered within Bell's classrooms, his house pupils became essential to his research, closely tying research to teaching. Bell became clearest and most emphatic about the relationship between his house pupils and his scientific work when the house pupils' place in his anatomy-school home was threatened by Bell's proposed marriage. In 1811, Charles was set to marry Marion Shaw, the sister of his brother George's wife, Barbara. Apparently, however, Marion objected to Bell's house pupils living with them. Charles Bell's response, conveyed to his brother George, demonstrates the value Bell placed on his house pupils, his teaching, and his scientific reputation: "I know that there is an ambition in my nature which cannot give up professional and scientific eminence—even to the attachment of a wife; and if by yielding now I lay a foundation for vain regrets, where is domestic peace? Now, my dear, dear George, you know how I love her . . . Be assured of this, that love of you and Marion is at the bottom of all my 'irritation.'"[66] There was an important issue at stake, after all. "You must consider that these young men are house-pupils, not boarders. Dr. Denman, Dr. Baillie, Mr. Hunter, Abernethy, Cooper, Wilson, have had them, or have them." This was a necessity for his professional ambitions. "It is no trick—no novelty,"[67] he concluded.

Bell's remarks neatly unite his zeal to attain professional and scientific eminence with the necessity of house pupils to the lecturer and, in so doing, tie that professional and scientific eminence to the classroom. He must do as other famous medical men and anatomists have done and take on house pupils, who are regarded as essential to the anatomy lecturer. Those house pupils were essential to the ambitious and scientifically minded lecturer because they helped to free up time by assisting with the lecture preparation; in addition, and crucially, anatomical research was conducted in and through the classroom, as Bell's work on the brain and nerves suggests. That union of teaching and research would eventually prove damaging for Bell, when one of his own particularly ambitious students, Herbert Mayo, claimed Bell's discovery for himself, but it was also an essential union for Bell, whose income and survival depended on classes that were developed through reputation, in part the reputation of the anatomist as a scientist, and one whose scientific eminence depended equally on his classroom success.

WRITING FOR THE ARTISTS AND NATURAL PHILOSOPHERS: *ESSAYS ON THE ANATOMY OF EXPRESSION IN PAINTING*

Charles Bell's written work in the period between 1804 and 1812—from his arrival in London to his purchase of the Great Windmill Street School—encompassed the subjects he taught and in which he hoped to achieve a reputation. When Bell left Edinburgh, he left with what was to become *Essays on the Anatomy of Expression in Painting* (1806) in hand. The book, like one of the courses he would teach, was meant for artists. The book was also meant for the philosophers in Edinburgh from whose company he had come. He set about working on it in his spare time immediately upon his arrival in London, as Bell himself recounted in the margins of his biography by Pettigrew: "I was not idle, however, all this time. I had a subject of study always with me. I was preparing my Anatomy of Expression, and made some anatomical studies in the Westminster Hospital."[68]

Preparation for the book and work on its subject took a variety of forms. In addition to anatomizing, Bell observed appropriate subjects, making observations in Bedlam and drawing from the patients, as in the instance in which he referred to having "a mad, laughing 'wife' to draw (if I can)."[69] But the sentence following suggests a broader understanding of madness, and indeed of the ambition of the book on expression, shaped by time in an Edinburgh milieu concerned with philosophy of mind: "I am reading Locke on Human Understanding, by which you may see that I am making a slow progression *back*wards. I wish to make a philosophical and methodical introduction to my Passions."[70] The book's intention, as declared in its introduction, was "to demonstrate the importance and the uses of anatomy; to multiply the motives for the cultivation of the science; to show how various and how interesting are the deductions which may be drawn from the contemplation of the animal frame."[71] Bell was writing about placing anatomy within broader contexts, following in the footsteps of William Cheselden and John Hunter in making it a subject for natural philosophy and not just professional training.

Evidence of Bell's natural philosophical aspirations (or perhaps pretensions) for the book can be found in how he described his work to his brother. In one letter he wrote, "I have made an observation,—I call it a discovery,—in studying the lions and tigers, that has greatly delighted me."[72] The discovery, in this instance, had to do with animals and mad men both having a smooth brow in moments of rage (rage being indicated by the lips and eyes in those cases), while in men among whom there was "some mental operation mixed with the passion, there is knitting of the eyebrows, and the peculiar conformation of the lips from the orbicular muscle—muscles which the animals have not."[73] But the point is less to do with the substance of the discovery than with Bell's self-conscious fashioning of an observation *as* a "discovery," a fashioning that he himself described in a letter.

Always eager to be a gentleman of science—someone with wealth and standing, like Humphry Davy or Joseph Banks—Bell talked in ways that suggested that he was hoping for greatness from this book.[74] In another letter, Bell wrote of his book on the anatomy of expression: "the subject everywhere admits and requires theory and reasoning . . . We shall be able to combine it [the book's 'insulated remarks'] in system, and then it will admit being talked of. If my sketches are true, my theory will not alter them."[75] Theory and reason, it seems, would save the book from being just a series of observations or sketches and would make Bell's observations into a system, such that the book would "admit being talked of." In this sense, the book was not particularly a practical textbook—it was a work of natural philosophy. And the words "system" and "theory" with which Bell described *The Anatomy of Expression* are very much the same sorts of words he would later use to describe his most famous discovery to do with motor and sensory nerves. *System* and *theory,* which seem to suggest ordering, context, and philosophical underpinning for Bell, are the elements that saved a work from being merely practical and professional—they were traits that gave something standing as natural philosophy. But natural philosophy was not a course of study for medical men and it did not pay bills, so the book needed to fill a second purpose, and it did.

It seems that, as he had hoped, Bell's book on expression was discussed. Bell's notes to his brother about its success include a reference to the book being "noticed in the Parisian journals," where they evidently said it was "the result of many years' study, and that the best critics speak highly in its praise."[76] Bell also wrote about a social gathering with John Murray, the publisher, and others, after which "there came a gentleman down-stairs in a great hurry, with Murray at his back, holding out his arms to me. He said he did not know that he had been in company so long with the author of the 'Anatomy of Expression.' This was Prince Hoare."[77] Prince Hoare, the English artist, leads us to the other purpose and audience for this broadly situated book of anatomy. *Essays on the Anatomy of Expression in Painting* was explicitly intended for artists, though it is interesting to note that Bell had the book reprinted in 1824 under the title *Essays on the Anatomy and Philosophy of Expressions,* suggesting that by that time he had a new target audience (natural philosophers rather than painters) in mind.

Bell began the introduction to the book with a description of his own professional training in anatomy as a surgeon, saying that while he was trained to look on the body as a medical man, he had always also seen the anatomy of the human body from another perspective as well—that of an artist. "Anatomy stands related to the arts of design," Bell says, "as the grammar of that language in which they address us. The expressions, attitudes, and movements of the human figure, are the characters of this language; which is adapted to convey the effect of historical narration, as well as to show the Working of human passion, and give the most striking and lively indications of intellectual power and energy" (figure 1).[78] It seems

FIGURE 1. Engraving by Charles Bell in his *Essays on the Anatomy of Expression in Painting* (London: Longman, Hurst, Rees, and Orme, 1806), 142. This engraving, drawn by Bell himself, shows the human expression of "Fear mingled with wonder." It heads the section titled "Wonder, Astonishment, Fear, Terror, Horror, Despair." Bell labored, in the text, to display his artistic talents and also to draw philosophical distinctions between expressions on the faces of animals and those of human emotions and the human mind. Image courtesy of the Wellcome Library, London.

very much an appropriate book for a man who would eventually fill his classroom with art students and who already hoped to win the professorship of anatomy at the Royal Academy of Art.

Election to a professorship at the Royal Academy was a complicated political matter, requiring both social and professional patronage. Explaining the process to his brother, Bell wrote in 1807, "you ask me if the election for this professorship is likely to be soon. The man (Sheldon) is alive, but he cannot now lecture; a vacancy is not declared, nor likely to be soon. The voters are the Academy, forty in number. Those who have influence are his majesty, the Royal Family, nobility, and gentry: all who ever had their faces drawn, or are in the habit of considering themselves amateurs."[79] In anticipation of the vacancy that seemed sure to appear later that year, Bell began gathering his support. Some came from the natural philosophical quarters that formed part of the audience for *Essays on the Anatomy of Expression*, as with Sir Joseph Banks,[80] of whom Bell wrote, "Sir J. says I am the person best entitled to the situation, and if his name can be of any service he shall be happy, or call on those he can influence."[81] The medical men too lent their backing, even bowing out in deference to Bell's candidacy. "Abernethy has said that he had a strong desire of offering himself, but in consideration of my pretensions, waives. So Astley Cooper writes me as strong a letter. Mr. Wilson wished to stand, too, but Dr. Baillie would not support him, but me. These are the three great schools of London."[82] In these circles, Bell's work as an artist was already known. Astley Cooper, for instance, had asked Bell to draw for his own work. Bell's early efforts at making acquaintances and friends of those whom he hoped would provide professional patronage had paid dividends.

The book, however, was especially useful in procuring social patronage—patronage from those outside his profession. It made quite an appropriate gift for royals whose support Bell courted, and it helped introduce him to artists.[83] Dr. William George Maton, one of Bell's friends, presented a copy of the *Anatomy of Expression* to Princess Elizabeth, presumably in the hope of securing royal patronage as well. According to Bell's letter to his brother,

> She was delighted with it . . . and *she was pleased to say it was* QUITE IN HER WAY! Maton then said I had a desire of presenting it also to her Majesty. 'Oh,' says she, 'have you a copy for the Queen?' So luckily he had, and she carried it to her mother. The Queen spoke to Maton of it on the birthday, with great admiration, and begged him to convey her thanks to me. One of the ladies of the bedchamber said to Maton the Queen was reading it for two hours last night. Oh, happiness in the extreme, that I should ever write anything fit to be dirtied by her snuffy fingers!'[84]

But despite being "backed by powerful friends,"[85] including the princess, Bell's medical patrons, and some of the painters, Bell lost the canvass to Carlisle, who was well connected and believed that students of art and sculpture did not need to know anatomy.[86] So Bell "resolved to set about [his] private class according to promise,"[87] holding private classes on anatomy for artists. Benjamin Haydon and others of the Royal Academy like the Landseers (a family of engravers and painters, who were well-recognized in their day) took anatomy from Bell because he taught hands-on dissection to artists, something others in London were not doing.[88]

It is notable that the first of his three books written during the period, *The Anatomy of Expression in Painting*, served the role of appealing to one of Bell's three core groups of students—his painters—as did each of his other books as well. *The Anatomy of Expression*, like the later *New Anatomy of the Brain*, is attentive to self-presentation and adoption of the rhetoric of natural philosophers, carefully framing overarching questions as philosophical (the former to do with questions of beauty and of human and animal minds; the latter to do with relations between the parts of the mind and nerves, between sensation and action), and its publication was meant to secure a place for Bell at the Royal Academy of Art. If that purpose went unfulfilled, it was not for failure to circulate the text, even to the Queen herself. Bell's book gained esteem, but his classes, while also popular, continued to leave him with debts well after 1806 and with no prospect of a regular salary or career path in sight. His next book would be written for a different audience, therefore, to fill another of his classes.

WRITING FOR THE SURGEONS: *A SYSTEM OF OPERATIVE SURGERY*

In 1807 and 1809, Bell published two volumes of a book to do with another of his regular courses: *A System of Operative Surgery*. Though he had not yet obtained a hospital position, and Bell's letters home say relatively little about surgeries or patients (practice for its own sake never being Bell's primary interest or path to success),[89] once Bell set up his little school, he taught a class on surgery in the evening. This book would be for them, his classes of surgeons.

The most practically oriented of Bell's works from this period, the book, he declared in a later (1814) introduction, was limited in scope and aim. Students, he said, must learn first from teachers in the hospital schools and anatomy theaters before they turn to books. He described his intended reader "as having attained a knowledge of Anatomy and of the Doctrines and Practice of Surgery." This was knowledge that could not be conveyed by books: "it is to be acquired only by continual exercise, by daily and careful observation, by treasuring up the lessons . . . drawn from his teachers."[90] The book was not meant to replace teachers and classroom experience, certainly not to replace Bell himself, since, drawing his income from his classes, the last thing he wanted to do was to render them redundant. Instead, the book was meant to *accompany* those classes—"a book in the hands of the pupil to direct him in his studies; to be associated with all he sees and hears; in which the lessons he has detailed to him at length by his teachers may be found more shortly expressed; to which, as a student, he can recur for a concise exposition of the points material in practice."[91] The book was also to be a handy reference for practicing surgeons, to save them the embarrassment of trying to remember what they should do or of having to whisper questions to others in front of the patient (as Bell himself puts it) when the time to operate came. Bell wrote that the volume was one "to which, as a surgeon, he can turn for the detail of what is necessary to be done in preparing for an operation and in operating, unembarrassed by useless disquisition."[92] It was, therefore, a practical book, a sort of reference work, in a way that *The Anatomy of Expression* was not.

Bell had acquired experience in the practice of surgery in his apprentice days serving under his brother John in Edinburgh, but he continued to accumulate observations to put into this book of common operations. In 1805, he wrote, "It seems as if of necessity I will soon become rich in cases of surgery if I have but common industry or management. I make a practice of noting down my observations as the opportunity occurs."[93] It was in that year, 1805, that Bell received an offer to publish the as-yet-unwritten book on surgery, for a sum that Bell thought quite handsome: "300*l.* for two 8vo. Vols," an offer that may have been the initial impetus for writing a book that Bell sometimes found tedious. Bell wrote to George with relief, "Now, my dear George, if—as I am afraid—I have led you into difficulties, this may at once extricate us, and be pretty well."[94] He accepted the offer, of course, but Bell found the writing of the book oppressive. The objects of the book—to fill his classrooms as well as to receive the publisher's sum—were clear enough to Bell, as was the need to achieve those objects. In 1806, he wrote to George: "Horner has my last essay in the proofs, for these three days, in his hand. I am now in a situation to profit by my works, and I think my Surgery will do me great good—in fact, establish me in practice."[95] But he wrote his brother in 1807, "You ask me to send the book—do you mean the book of surgery? The proof by me is at page 144, and I have all the rest to write. It presses me like a nightmare."[96] Still, the work did what it was meant to, and Bell's classroom benefited.

In 1807, the year in which the first volume of his treatise was published, Bell wrote to his brother that his course on surgery was "overflowing . . . absurdly full," even at the same time that his lectures on anatomy were faltering. If teaching lay at the center of Bell's attempts to earn enough money to survive and perhaps even to fashion a career, his written work cannot be seen as separate from that endeavor. As with *The Anatomy of Expression*, used to recruit supporters for a canvass and students to the classroom, *A System of Operative Surgery* functioned well as a promotional tool for its author and his classes.

Bell continued to work on his treatise on surgery even after the first edition came out, developing practical skills and planning for a second edition. His primary new contribution to the second edition was to be an illustrated section on gunshot wounds, though as he wrote to George, "the booksellers wish it to be published separately."[97] That appendix of illustrations, along with "noble specimens of injured bones, and a series of cases admirably fitted for lectures,"[98] came from Bell's experience with wounded soldiers at Portsmouth after the Battle of Corunna in 1809. In Haslar Hospital until eleven at night, by his own account, Bell came away from Portsmouth with respect and fondness for British soldiers, cynicism about their military leaders, and a wealth of new experiences and new specimens to inform his classroom lectures, illustrations, and museum. He was touched by what he saw there, writing to George about a scene of the sort that he hoped George himself would never see, "I have stooped over hundreds of wretches in the most striking variety of woe and misery, picking out the wounded. Each day as I awake, still I see the long line of sick and lame slowly moving from the beach: it seems to have no end. There is something in the interrupted and very slow motion of these distant objects singularly affecting."[99] He also learned a great deal and considered it an important part of his own ongoing training as a surgeon, writing "shame upon the fellows who did not take my example to learn their profession."[100]

The wounded soldiers also provided an excellent source for Bell's museum, an important part of his pedagogical program. He made sketches for his book or museum (their home not yet certain) and then worked on etchings of them, teaching himself the technique. He wrote George, "I am so much improved in my etching, that I have rejected my first plates. I have bought etchings by the best old masters, and have hung them around me; they will prevent my being little and feeble."[101] He also painted the soldiers using watercolors, and those paintings, themselves full of pathos and a tragic beauty, illustrated suffering and surgical stories (figure 2). They were placed in the museum, "in the interstices of the preparations."[102] He also took "noble specimens of bones." He followed up with the men in London after they returned, writing to George, "Went also to Chelsea to converse with my gun-shot men; expect thirty-two of the Corunna army to-morrow. I have made out a great deal of principle, and have cases, or rather instances, to illustrate every point."[103] His patients became illustrations in his museum and book, illustrations meant, of course, for students.

As with Bell's natural philosophical writing, his practical, surgical work was seldom much separated from his classroom work. Teaching was again at the center. And Bell was popular. "You ask me of my pupils," he wrote, "they are very fond of me. One fellow I thought long my enemy, but I found

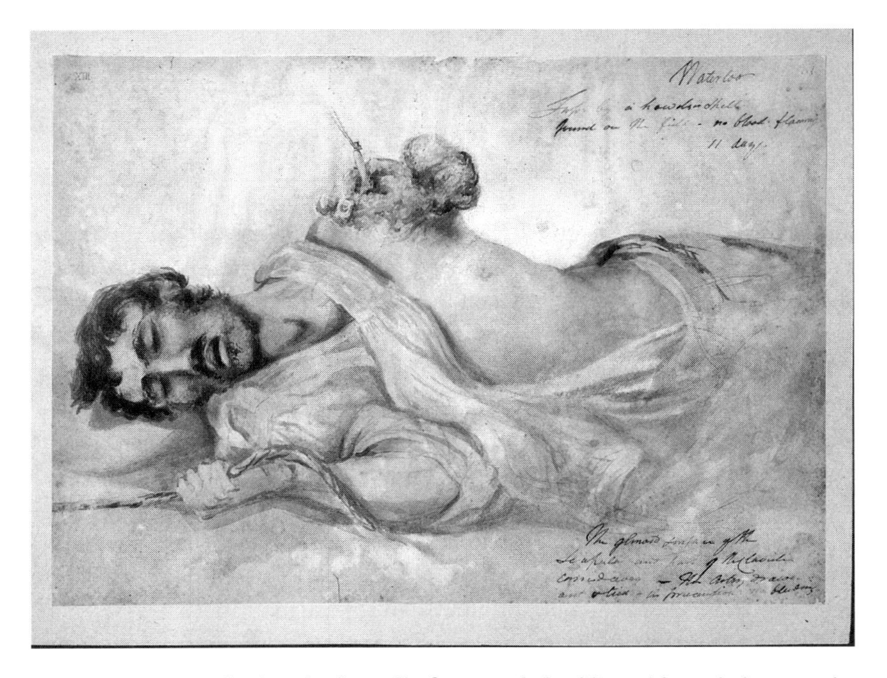

FIGURE 2. Watercolor by Charles Bell of a wounded soldier with a missing arm, inscribed "XIII, Waterloo. . . ." This is one of Bell's watercolors made from his sketches of wounded soldiers from the Battle of Waterloo whom he treated in Brussels in June 1815. Such paintings were made for teaching purposes. From the Royal Army Medical Corps Muniment Collection. Image courtesy of the Wellcome Library, London.

that he, more than any other, had been beating up for recruits for my class next season."[104] But Bell sought a reputation and better standing as well. He wrote George, "As it is, I sometimes strut, sometimes hang the head. I can only study when my pockets are full, which I conceive to be a demonstration that I was intended for a gentleman."[105] Though his classes were not yet as successful as he intended them to become, and the status of "gentleman" would not yet come from full pockets, Bell sought natural philosophical work that would give him standing in society. In 1807 he wrote, "I am casting about for a subject to make something new of. I have been thinking of the brain—of mind—of madness. Could I not put this subject in the form of queries as to the best way of prosecuting the subject, to be laid before Stewart or Jeffry, &c.[106] I would not publish any thing but in papers for this many years."[107] That work on the brain, which Bell himself conceived of and fashioned as "something new," much as he had done with *Essays on the Anatomy of Expression in Painting*, would occupy the rest of his career, both within the classroom and beyond it.

WRITING FOR ANATOMISTS, NATURAL PHILOSOPHERS, AND FOR FAME: *IDEA OF A NEW ANATOMY OF THE BRAIN*

By setting out to write on the brain and nerves, Charles Bell was taking up a topic that had been central in Edinburgh, where phrenology was to reign, and also where philosophical and anatomical discussions of the mind, the brain, and the senses had long occupied a place of significance (figure 3).[108] The selection of a topic with that kind of prominence in Bell's world was not accidental—new work in that area would likely make a name for someone, and that was what Bell had in mind. In December 1807, he wrote to his brother, "My new Anatomy of the Brain is a thing that occupies my head almost entirely. I hinted to you formerly that I was *burning*, or on the eve of a grand discovery."[109] His grand discovery would ultimately have to do with identifying separate roots of motor and sensory nerves, but his claims to scientific innovation would be contested for much of Bell's life.

At this point, however, it was something less precise, something captured more by the idea of a system—the idea that the nerves were organized in a symmetrical, circulatory fashion—than by a discrete anatomical fact, like the discovery of a new organ or its function. Bell established this exact point in a letter to his brother, saying, "I have completed my view of the Brain, but it is only the introduction to the strict anatomy, giving a view of my system, for I find that it embraces the whole nervous system."[110]

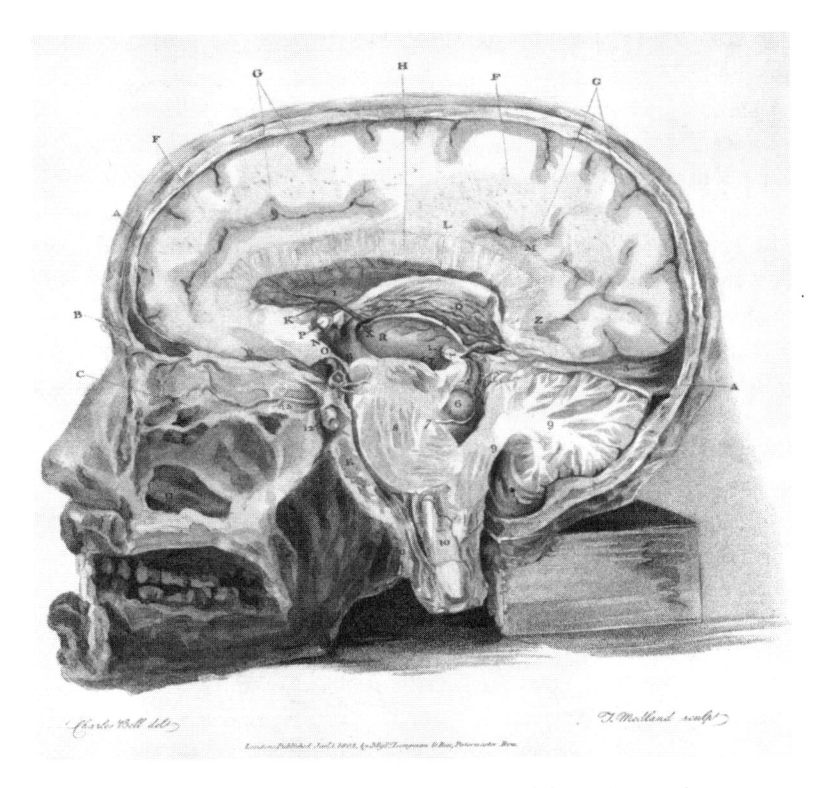

FIGURE 3. Drawing by Charles Bell for his *Anatomy of the Brain* (London: Longman, Hurst, Rees, and Orme, 1802), facing p. 41. The image shows a section of the skull "cut a little to the left of the course of the longitudinal linus." It is representative of Bell's early anatomical (dissection based) work on the brain. Image courtesy of the Wellcome Library, London.

One striking element of Bell's work was his reliance on "systems" as characterizing good science. Bell talked about systems of nerves as a way of implying the interconnected, elegant, completeness of his discovery, and about systems of medical education with much the same set of characteristics in mind. Such talk is particularly interesting in the face of the relatively common contemporary rejection of "systems," a term usually associated with theoretical medicine, in favor of empiricism in French and American, as well as British, medicine.[111] Bell was certainly an empiricist and considered empirical medicine to belong properly to Britain, so what then did he mean by "system"? The answer to that question seems to have been shaped in Edinburgh and can be found in part by looking to the work of William Cullen, which infused Edinburgh's medical curriculum and serves to caution historians who would like to see system and empiricism as antithetical.[112]

As Christopher Lawrence has described it, William Cullen used the nervous system as a basis for teaching medicine to students systematically.[113] The mechanisms by which the mind worked, according to Cullen's friends Adam Smith and David Hume, required that all natural philosophy be taught systematically in order for the mind to comprehend and learn it. Cullen therefore built a system of medical training and of disease taxonomy (nosology) around the all-important nervous system, stressing theory, system, and simplicity in his medical and chemical lectures, much as Bell would later.[114] His views would have predominated among those trained in Edinburgh and those who continued to teach in Edinburgh and in London well into Bell's time. Lawrence writes, "Cullen began with a history of medicine, followed by a discourse on the method of studying medicine, and then an account of nosology. Here Cullen produced an extended defense of theory and system in medicine founded on the premises either drawn from, or shared with David Hume and Adam Smith."[115] While such systems were theoretical and speculative (built that way deliberately for pedagogical purposes, for the world was ordered and made sense of by the human mind using causal relationships), Cullen was also "intensely empirical" at the bedside, seeing appropriate roles for system, skepticism, and empiricism simultaneously in the development of natural knowledge.

For Cullen's friend Adam Smith, who used the word "system" often, a system seemed to mean something comprehensive, with no gaps, coherent, and simple. Smith used his history of astronomy to discuss the ways in which the mind seeks pleasure in order, whereas disorder produces first wonder and later anxiety (hence the long history of efforts to use theories to order or make systems out of observed astronomical phenomena). Smith also discussed the matter of system early in "Letter to the Authors of the Edinburgh Review" (1756), and in other early essays, as well as in *Theory of Moral Sentiments* and *Wealth of Nations*.[116] "System" seems to have meant something similar to Bell, who used the term often to espouse the virtues of his work when compared to those of experimentalists bent on establishing facts. In one such example, Bell wrote to his brother, "I think the observations I have been able to make furnish the materials of a grand *system* which is to revolutionise all we know of this part of anatomy."[117] Dugald Stewart, Smith's student, wrote that Smith described the rise of a "systematical beauty which we admire in the structure of a cultivated language."[118] The pairing of the systematic and the beautiful is one that was echoed in Bell's own work. Systems—whether of anatomy or of language—were orderly, beautiful, comprehensive, simple, and elegant, and they fitted with Bell's Edinburgh upbringing among the philosophers of the Scottish Enlightenment. They were not in any sense antiempiricist; they just made discrete facts into something that mattered and, more importantly, something that could be learned.

The "system" (a loaded term, to be sure, in Bell's day[119]) developed here brought together the brain and the nervous system, the mind and the senses, and established relationships and connections. In 1807, in the early stages of his work, here is how Bell described his "grand discovery":

> I consider the organs of the outward senses as forming a distinct class of nerves from the other. I trace them to corresponding parts of the brain totally distinct from the origins of the others. I take five tubercles within the brain as the internal sense. I trace the nerves of the nose, eye, ear, and tongue to these. Here I see established connections. Then the great mass of the brain receives processes from these central tubercles. Again the great mass of the cerebrum sends down processes or crura, which give off all the common nerves of voluntary motion, &c. I establish thus a kind of circulation, as it were. In this inquiry I describe many new connections. The whole opens up in a new and simple light; the nerves take a simple arrangement; the parts have appropriate nerves; and the whole accords with the phenomena of the pathology, and is supported by interesting views.[120]

What is noteworthy about the passage is that Bell's anatomy is a sort of philosophical anatomy, though not the sort of later transcendentalist philosophical anatomists like Lorenz Oken and Étienne Geoffroy Saint-Hilaire.[121] Neither was his contribution that of a purely descriptive, technical anatomist of the sort that abounded in London during the period.[122] Instead, he was a natural philosopher, looking for systems, functions, and connections in anatomy, and asserting for anatomy a right to stand as a properly philosophical subject within natural philosophy. While in practice his anatomy took a rather Cuvierian tone—pairing form with function in law-like fashion[123]—Bell's work on the brain also established connections, distinctions, and analogies, relating various nerve roots and ganglia to each other, but also to the system of the circulation of the blood (not dissimilar to Cullen's analogy between the nervous system and electricity), reasoning by analogy that nature would likely work in a symmetrical and circulatory fashion with nervous impulses as well.[124]

In this instance Bell labeled distinct sets of nerves for the outward senses as corresponding to distinct portions of the brain. Five tubercles within the brain receive the nerves conveying sight, touch, smell, taste, and sound, and the "great mass" of the brain receives impulses from these five tubercles and sends signals for voluntary motions to the limbs. Such an imposition of a kind of structure seems, at this point in Bell's investigation, to be entirely based on ideas about how bodies *should be* divided into structures, though it still seems to be governed by the rule that "form follows function." The purpose of Bell's dividing and connecting of anatomical parts seems to have been an analogy of the sort that made his work appear properly natural philosophical and that would make his discovery one of high stakes, whose discoverer would be worthy of fame and fortune. It was an analogy that suggested that the functions of bodies are governed by some kind of unity—the nerves, like the blood, form a *circulatory system*, one for which the brain is the central hub. This "great discovery" was hardly one of a descriptive anatomist; it was meant to be philosophical. Like Bell's politics, his anatomy straddled and encompassed various approaches, styles, and seemingly opposed positions

As his discovery took shape, Bell reluctantly turned to another method of proof not born of descriptive anatomy: to describe parts of the brain that were distinct in function by tracing them outward through the nerves, he felt he had to perform vivisection experiments on donkeys. By 1810, three years after Bell had come up with the basic analogy of circulation that defined his system, he had settled on something a bit more precise, with a form of inquiry appropriate to it. He hypothesized that divisions of the brain corresponded to divisions of the spinal marrow and that nerves contained within the spinal marrow each had two roots, bundled together. The roots of the spinal nerves, then, could provide access to the brain and its functions through experimentation.[125] The experiments were both difficult to prosecute and morally repugnant to Bell,[126] and yet they promised to yield "facts the most important that have been discovered in the history of the science."[127] For the initial experiment, he opened the spine of an ass and pricked and injured the posterior filaments of the nerves. When he saw no motion in the muscles, he pricked the anterior division and saw that the parts convulsed immediately. What Bell was really after, in pricking the roots of a living animal's spinal nerves, was the functions of the parts of the brain, an organ that offered little in the way of distinct anatomical parts subject to examination. The second experiment built on the first. Bell wrote: "I now destroyed the posterior part of the spinal marrow by the point of a needle—no convulsive movement followed. I injured the anterior part, and the animal was convulsed."[128] The result of the two experiments was obvious to him: "the part of the spinal marrow having sensibility comes from the cerebrum" (what Bell had earlier in the letter called the "organ of mind"), whereas "the posterior and insensible part of the spinal marrow belongs to the cerebellum."[129] The discovery and its changing contours and contests will be the subject of chapter 5 of this book, but this is where things stood for Bell in 1810, just before the 1811 printing of his *Idea of a New Anatomy of the Brain*. He already saw it as the work of a lifetime, writing that he would continue to develop in the years to come.[130] But if the discovery was what Bell thought, he wrote to his brother, "I am made, and a real gratification ensured for a large portion of my existence."[131]

This circulatory system of the nerves, at whose center stood an inquiry into the brain (the organ that most concerned the Scots in Edinburgh), occupied Bell almost entirely for much of 1807–11. He recounted nights of little sleep spent thinking about it, saying: "But last night I took a long pull at the subject of my most anxious contemplation, 'The Brain,' and so heated myself with it, that at half-past two I had no more disposition to sleep than now."[132] The book was meant to be a little thing, just to test Bell's idea and to make sure that the most important natural philosophers and medical men knew about it. It was not meant to explain the anatomy of the brain to novices—that would come later, and in his classes—instead, it was meant to impose a system and organization on what was already known about the anatomy of the brain and nerves, for those who already knew something of those anatomical systems.[133] After the book would come a series of papers for the Royal Society laying out the system in the first, its detailed anatomy in the second paper, and its pathology in the third.[134]

Charles's brother George would be paying for his little treatise, like much else Charles produced. In 1809, Charles wrote to George, "Speaking of books, could you get a little tiny book printed for me, of twenty pages of the smallest 12mo [duodecimo]? For I must send you down the manuscript of the Brain again, stated shortly for my friends."[135] In the end, the little book was printed in London, in 1811. Bell specified often in letters those individuals who would constitute the initial audience for the manuscript and whose opinions he sought: Francis Jeffrey, Dugald Stewart, and Henry Brougham.[136] That list, together with a list of those who should receive copies of the printed book that Bell drew up for his assistant, John Shaw, encompassed Bell's patrons, medical and social, and included names we have seen before—Abernethy, Baillie, Cline, Cooper, Gartshore, Horner, Maton, and Wilson—as well as some that are familiar for their success as natural philosophers—Humphry Davy, William Hyde Wollaston, Peter Mark Roget, and John Playfair. Twenty copies were simply listed as going to Edinburgh.[137] Bell's ambitions for the little book are clearly spelled out in that list of the most famous philosophers and medical men of the period. He imagined its success in a letter to his brother, saying, "I think to the profession at large it will prove most acceptable; and while some will adopt it, I trust the most captious will say it is ingenious."[138]

The brain occupied Bell's daytime hours, too; he allowed it to, because, like everything he worked on, Bell incorporated this work into his teaching. Of what became the "little book," *Idea of a New Anatomy of the Brain*, Bell wrote, "The Brain I wish still to resume, after giving out a short account of my view as taken from my lectures. It was this which I proposed you to print in Edinburgh."[139] The content of his great discovery, in other words, was built in the classroom, through lectures. In London's competitive medical marketplace, such a discovery, developed by one whose central occupation was teaching, would be publicized through the classroom as well, where it could be used simultaneously to bring him fame and, by doing so, draw students to his classes, bringing him fortune as well. Bell said of his plan for the discovery: "My object is not to publish this, but to lecture it—to lecture to my friends—to lecture to Sir Jos. Banks' coterie, to make the town ring with it, as it is the only new thing that has appeared in anatomy since the days of Hunter."[140] In talking about "Sir Jos. Banks' coterie," Bell was clearly trying to build his professional circle, referring to a group of natural philosophers and not just medical men, hoping to engage that wider group in his classroom. It was a book that was self-consciously fashioned, much in the same way Bell's networks were.

In three very different books—one meant to be a philosophical text uniting anatomy and dissection with the fine arts, one meant to serve as a supplemental quick reference for the surgical student or trained surgeon, and one meant to convey a natural philosophical system constituting a novel discovery—Bell's classroom was always at the center. In his *Essays on the Anatomy of Expression in Painting*, Bell made a first attempt to build an audience that his classes on fine art and anatomy catered to and to make social, political, and professional connections that might lead to a professorship at the Royal Academy. Bell strategically introduced his second book, *A System of Operative Surgery*, as a companion to classroom lectures, and he filled his classes on surgery with students the year that the book came out. The third book in the series corresponded to the third and final set of classes he taught—classes on anatomy for anatomists and natural philosophers. Through those classes, Bell aimed to position himself as a scientific thinker of distinction, one responsible for an important anatomical discovery, and his subject, anatomy, as one proper to natural philosophy. *Idea of the New Anatomy of the Brain* was written more about a philosophical *system* of anatomy than about detailed observations on anatomical structure. It was written for Edinburgh philosophical luminaries like Dugald Stewart and London's natural philosophy elite, such as Joseph Banks and Humphry Davy, and it was drawn from Bell's lectures and was publicized through them as well. *Idea of the New Anatomy* was part of Bell's self-presentation as a natural philosopher and his claim to fame on the London scene.

THE FLEXIBLE POLITICS OF A PROFESSION IN THE MAKING

It was not clear, in 1800, how one built a career as a natural philosopher and anatomist in London. Instead, careers were cobbled together by assembling collections of patients, paying students, and patrons. Charles Bell particularly depended on his classes to make a living—the source of income that corresponded best to his natural philosophical and gentlemanly ambitions. He was rather successful. By November 1808, he was able to write to his brother, "my Classes are likely to be popular . . . they are increasing . . . my whole exertion must be in that direction: and having entirely succeeded, then it is like one of Buonaparte's battles—the lesser circumstances will arrange themselves accordingly."[141] Even success, however, was relative and did not fill the coffers. He described his evening lecture, which he described as "stuffed to suffocation" with thirty-six pupils,[142] as a good start at least: "you must recollect," he told George, "under what disadvantageous circumstances I began." His competitors, after all, were doing no better: "You must take it into consideration that Carlyle is lecturing to four pupils: that Thomas, who once had a good class in the next street, has knocked under; that Chevalier does not lecture, and that Pearson does not consider it advisable to lecture this year; that Homer gives gratis lectures to the students of St. George's Hospital—twenty-five in number." He concluded with some satisfaction: "An overflowing class, then, of thirty-six is not to be grinned at."[143] Bell's books, reputation-building ventures in their own right and meant to establish Bell's place among natural philosophers and the professional elite, were also meant to promote Bell's classes and to help them continue to grow. Bell's pursuits were all intertwined.

A biographically focused examination of Charles Bell's early years in London must cover his initial attempts to build networks of patronage, his early publishing ventures, his classes, and his attempts to claim a professorship; what unites those various pursuits is in fact that which helps to shed light on London medical science during the early nineteenth century more generally—absent a clear career path, scientific medical men cobbled together a living from a variety of occupations and endeavors in the hope of gaining a reputation and a stable professional position. This sort of social system necessitated flexible politics of the sort that Bell embraced.

Reflecting on his career to date in 1811, at the end of the period under consideration in this chapter, Charles Bell wrote to his brother George, "I was too early brought forward as a teacher, and too much left to my own weak efforts, not to feel acutely what I went through on leaving Edinburgh. The length of time which I saw must pass before I could possibly accomplish what I was resolved upon—a proud station in the profession,—threw me off, as it were."[144] Professional accomplishment and standing were developed over years, and not by following a predictable and well-trodden path. Bell's cultivation of a career and his deliberate self-fashioning as a gentleman of science involved developing robust and wide-ranging social and professional networks. These networks depended on the kind of flexible politics that could fit in anywhere, as well as appropriate manners, outfits, and lodgings, and the work of calling on friends and colleagues and dining with their families, friends, and patrons.

Still, Bell was not without commitments. His unwavering fidelity was to teaching as the basis of a science of anatomy. As a scientific calling, teaching needed to be systematic, and accordingly, Bell wanted to found a school for comprehensive training in anatomy, surgery, and medicine. It was his great goal. In the reflective letter to his brother quoted above, Bell wrote, "I am very desirous to have my class-rooms removed from this place [Leicester Square], and to found a school. If I am baulked here, I have no object remaining but the advancement of John Shaw. I shall continue to give lectures until my patients increase sufficiently [to] give it up, only retaining my lecture-room to give lectures on some chosen subjects occasionally."[145] Teaching, unlike Bell's politics or his other social endeavors, was not *simply* a means by which to make a living. His commitment to it was idealistic and total. Should he not be able to achieve his aims and found a school that implemented his pedagogical philosophy, he was prepared to walk away, to resort to practice instead. The founding of such a school would remain his ambition through the 1830s, with his successive attempts to build such a regular medical education for students and a career path for teachers at the Great Windmill Street School first, London University next, and finally the

Middlesex Hospital School. But already in the first decade of the nineteenth century, his first years in London, Bell recognized, in a forward-looking passage: "My means of being known are through my books and pupils: I retain my consequence by preferring science to practice. My chief gratification is in the cultivation of my profession, and I have still some great schemes to be brought forward."[146]

CHAPTER TWO

Pedagogy Inside and Outside the Medical Classroom: Training the Hand and Eye to Know

> When I have a subject that admits of reasoning and deduction, I have my applause. Winn, who is a gentleman and a scholar, said that he has seldom enjoyed such uninterrupted delight as in hearing my lecture, on the skull, to the painters. I find in my lecture that the only thing worthy of preparation is arrangement—and instead of a catchword, a leading idea. I could put the notes for a lecture on my nail.
>
> CHARLES BELL TO GEORGE BELL, February 8, 1808[1]

In 1811, Charles Bell was happily lecturing to increasingly large crowds in his own little house at Leicester Square. By 1812, he had married Marion Shaw and moved his residence to Soho Square (the same square that was home to Joseph Banks, Henry Fuseli, and others among Bell's artist and natural philosopher friends).[2] Soho was also home to the Great Windmill Street School, which became the focus of Bell's ambition. Early in 1812, James Wilson, then proprietor of the Great Windmill Street School, offered to sell the school to Bell for ten thousand pounds. It was a price Bell could not afford. Instead, he agreed to pay two thousand pounds to Wilson in order to be named proprietor, but allowed Wilson to continue to lecture at the school, to reside there, and to take house pupils.[3] George was not happy with Charles's arrangement, but at the Great Windmill Street School, Bell's museum and classes flourished. Bell's classes were full, with eighty to one hundred twenty students—bigger crowds than he had ever had before—and he found himself "at the height of [his] ambition in regard to teaching."[4]

Perhaps as importantly for Charles, the Great Windmill Street School gave him a place among a fine lineage of anatomists and a reason to suppose that, as an heir to the best of Scottish anatomy, he too was destined for success and fame.[5] The Great Windmill Street School, founded by William Hunter in 1767, was home to a line of distinguished anatomists, most of them Scotsmen. Hunter began the Great Windmill Street School after his offer to Lord Grenville to establish a national school of anatomy and an anatomy museum was declined. He had bought, collected, and prepared an extensive collection of normal and morbid specimens to fill such a museum, and those specimens found a home in the Great Windmill Street School, which Hunter had built with a museum, a theater, and dissecting rooms. Following Hunter's death in 1783, the Edinburgh-trained William Hewson took over, followed by William Cruikshank, an Edinburgh native who had attended the lectures of both John and William Hunter. William Cruikshank was succeeded by the Hunters' nephew, Matthew Baillie, a native of Lanarkshire, Scotland, who trained in Glasgow; then Baillie was succeeded by one of Cruikshank's pupils, James Wilson, who was from Ayrshire, Scotland, and had also studied under John Hunter. When Bell bought the Great Windmill Street School from Wilson, he had clearly bought himself a place in a school that had been kept within a sort of family of anatomists since its founding and had thereby secured himself a place within that "family" with the purchase.[6] If Bell's first years in London were spent fashioning himself as an anatomist and natural philosopher, and a gentleman deserving of a place in communities of surgeon anatomists and natural philosophers alike, his years at the Great Windmill Street were more focused—they were those of a man who had begun to feel that he had a place in the London scene and who now needed the finances and institutional arrangements to sustain it.

Bell began his time at Great Windmill Street in the museum, writing of the "happiness of this life of exertion, modelling [*sic*], writing, and putting this great museum in order."[7] The museum was full and, in a short time, so well developed that Everard Home (one of the previous teachers at the school and current president of the Royal College of Surgeons), as well as William Clift, the first conservator of the Hunterian Museum, gave their approval.[8] If the name of the Great Windmill Street School did not do enough to position Bell within a great historical school of anatomists, perhaps the idea was conveyed by the busts of William Hunter, Vesalius, and William Cheselden that Bell bought to ornament his museum gallery,[9] or by a copy of Joshua Reynolds's painting of John Hunter that hung there as well.[10] In any case, it was clear to Bell that he had made it.

William Brande (successor to Humphry Davy as the professor of chemistry at the Royal Institution[11]), James Wilson (the former proprietor of the school[12]), Benjamin Brodie (surgeon at St. George's Hospital and Fellow of the Royal Society[13]), and Peter Mark Roget (at that time known to Bell primarily as Sir Samuel Romilly's nephew[14]) came by the new school to talk to

Bell about advertisements and hours of attendance.[15] Ever conscious of his networks of patrons and of his own social standing, Bell wrote to George, "Now you find me united with a great body of men and beginning that connection with the character of sparing neither labour, nor time, nor expense for the advancement of the profession. Indeed I feel every day the advance I have made . . . I am, I may say without affectation, astonished that I have surmounted the difficulties and uncomfortableness that I have done."[16] The museum and school that Bell had bought came with a pedagogical legacy and not just a reputation.

William Hunter, who had laid down an educational philosophy in his printed introductory lectures, wrote about an interconnectedness between a museum of specimens and the classroom itself, a connection that was central for Bell during those Great Windmill Street years: "Large collections [of anatomical specimens] which modern Anatomists are striving, almost every where to procure, are of infinite service to the art; especially in the hands of teachers. They give students clear ideas about many things, which it is very essential to know, and yet which it is impossible that a teacher should be able to shew otherwise, were he ever so well supplied with fresh subjects."[17] Hunter was an important source of Bell's own teaching practices, and Bell noted with pride in an 1805 letter that the well-known surgeon William Lynn had said to him: "You must remain here; I see you must. I see you are calculated to be William Hunter amongst us."[18]

If Bell was calculated by Lynn to be the William Hunter of the nineteenth century, he seemed to be cultivating a reputation like that of William's brother John as well. John Hunter's reputation had been built and magnified by the Hunterian Orations of Bell's contemporaries, such as Everard Home, William Blizard, Henry Cline, and John Abernethy. As L. S. Jacyna has argued, the Orations, meant to preserve John Hunter's legacy, also served to elevate Hunter's reputation as the father of scientific surgery and of comparative anatomy and physiology in Britain (thereby proclaiming that surgery of their own age had already been transformed from a craft practice into a science).[19] By their account, the museum collections were Hunter's laboratory, showing system and order in anatomical parts, whether diseased or not. Such a portrait of John Hunter was intended to accomplish for surgery precisely what Bell hoped to do—to place it within the purview of legitimate scientific inquiry. Bell was surely aware of this image of Hunter, and perhaps he played to it himself, since the way he shaped his own identity seems to be very much like it. But, when Bell wrote of the Hunters, it was most often of William, and it is William Hunter's written work on the practice of teaching that promises to shed some light on Bell's practices as well.

William Hunter's introductory lectures provided a relatively comprehensive treatment of the ways in which training was meant to be conducted in the anatomy classroom. Bell's classroom, the same physical space as the one about which Hunter spoke (figure 4), was filled with many of the same objects, and almost entirely the same practices. But while Hunter wrote about the proper practice of teaching anatomy, Bell, taking the same material and joining it with educational philosophy as well as Cuvierian anatomy, crafted

FIGURE 4. "A lecture at the Hunterian Anatomy School, Great Windmill Street, London," watercolor by Robert Blemmel Schnebbelie (1830). This small (23 cm by 29 cm) watercolor depicts how the painter imagined the Great Windmill Street's famous anatomy school under William Hunter sixty years earlier. One can only imagine that such a scene was inspired by the Great Windmill Street Theatre during Bell's London years. Note the various objects and drawings arranged together in the classroom—a bust in the middle of the room, bones mounted on the chalkboard, a wet preparation on the lecturer's table, the skeleton suspended, and drawings all around. Image courtesy of the Wellcome Library, London.

from it a vision of a medical science suited to the sort of conservative reform that Bell favored.

If a student were to enroll in one of Charles Bell's anatomy classes at the Great Windmill Street School, he would find himself surrounded by objects, some of which would be familiar and some foreign to him. Those objects helped to constitute a pedagogical program at the center of Bell's medical science, in which surgery and general medical practice were taught through the cultivation of sensory perception and training of hand and eye, such that accumulated sensory experience could be, at more advanced stages, generalized and systematized. A system of display to teach the senses, largely built around the museum on which Bell worked so hard, provided the primary means of learning and of expressing knowledge in Bell's anatomy classrooms.

Students who entered Bell's classrooms watched Bell and his house pupil dissect a cadaver at the front of the room, over a full course, stretching weeks on end. They heard him narrate the dissection, clarifying the relationships between parts in a messy anatomical field, and they watched him impose order with rough sketches on the blackboard showing function or how parts came together. These same students could walk a short hallway to a museum full of collections of dry and wet specimens, collected to showcase multiple examples of the same part in order to give a sense of anatomical variations considered to be normal and those which could be classified as pathological.[20] Such specimens were sometimes brought into the classroom for lectures, where students would pass minute and detailed examples of anatomy not visible in the performance of dissection, or rarer specimens preserved because they were hard to come by. Bell's illustrious predecessor at the Great Windmill School, William Hunter, had described in his book of introductory lectures the way in which specimens were to be passed around the room, one student describing to the next what was to be seen. And students learned by doing—training their hand and eye together by preparing their own specimens and preservations, using models made of wax, and drawing from the specimens and from atlas illustrations. The hand and eye were seen as analogous organs, their training interrelated and simultaneous: doing was a part of seeing.[21] Reference books for students also served as manuals for practitioners. These were the material objects of a method of practical instruction.

In 1814, Charles Bell wrote to his brother that his "whole [pedagogical] system must be kept in full operation—preparations, drawings, models, cases, lectures, clinical lectures, &c."[22] It was, for him, a system, his anatomical science, and it was visual as well as pedagogical in nature. It was visual in the sense that all parts of the system Bell identified were visual and that seeing was primary and therefore privileged, but not in the sense that all elements of the system were *simply* visual; and the system was pedagogical because knowledge during Enlightenment and early nineteenth-century anatomy was produced and disseminated in a pedagogical setting.

TEACHING THE EYE TO SEE AND THE SENSES TO KNOW WITH SYSTEMS OF DISPLAY

Knowledge in late eighteenth- and early nineteenth-century anatomy was made and taught through a *system of display* of which these collections of specimens were only one part.[23] That system was used in pedagogical contexts, contexts that were also the sites of anatomical research. Anatomical museums functioned alongside books and bodies to form a set of instructional tools and objects of study. Individual elements of display, such as atlases or dissected bodies, were understood to function together as parts of a system that was created and deployed in a particular setting and for a particular purpose. That purpose and that setting were classroom based. To understand British medical science of this period, one must enter the classroom, see how various displays were used in practice, and then understand how they functioned as the basis of the science itself.[24]

The objects of study in anatomy included wax models; preserved specimens in jars, housed in collections; schematic chalk drawings from the classroom; elaborate engravings, and less elaborate etchings, found in books; dead bodies; paintings and sculptures; and living bodies. Together, these objects were taken to form a number of things: Nature herself, a representation of nature, and a representation of a particular argument about how the natural world was constituted. They were material embodiments of scientific knowledge at the same time as they were the anatomist's objects of study. As a constellation of interrelated tools, along with the catalogues and captions that framed them, they were understood to work together to serve both teaching and research functions, endeavors that often coalesced in a science that was rooted in the classroom.

William Hunter, Bell's much-admired predecessor at the Great Windmill Street School, discusses these in great detail. Hunter begins his posthumously published *Two Introductory Lectures, Delivered by Dr. William Hunter to His Last Course of Anatomical Lectures at His Theatre in Windmill Street* (1784) with a discussion of the origins of anatomy, saying that "the observance of bodies killed by violence, attention to wounded men, and to many diseases, the various ways of putting criminals to death, the funeral ceremonies, and a variety of such things . . . have shewn men, every day, more and more of themselves; especially as curiosity and self-love would urge them powerfully to observation and reflection."[25] The sort of casual and unstructured, informal experience of death as well as of living bodies formed one end of the spectrum of displays that were understood to constitute the experience of anatomy students and medical students. Funerals, executions, deaths, and illness, human bodies and corpses were not foreign or distant things to medical men in Georgian Britain. In such encounters—visual displays of death, a theater of dead bodies—the bodies themselves were not objects of formal study, but, as Hunter himself articulates, they functioned as a backdrop, a set of common experiences that would underlie formal anatomical study. Such bodies, unlike the corpses of the anatomical theater, were not especially the objects of scientific intervention or the subject of a particular scientific aesthetic; their uses were not understood to be scientific, and their audiences were multiple, but they served a role in a pedagogical philosophy that foregrounded experience with material objects as the foundation of knowledge.[26]

Such casual experiences with dead and diseased bodies were complemented by formal dissections taking place in lecture theaters. John Bell, Charles's oldest brother, said of dissection that it "is the first and last business of the student."[27] William Hunter had the Great Windmill Street School's theater built with the care befitting such important instructional work, saying:

> You may observe that this theatre is particularly well constructed, both for seeing and hearing; a strong sky-light is thrown upon the table, and the glass being ground, that is, made rough upon one surface, the glare of sun-shine is not admitted: the circular seats are brought as near the table, as ease in sitting would admit of; and, as they go back, they are a good deal raised, which is a considerable advantage both in seeing and hearing. . . . [T]he table, where the object is placed, and by which the demonstrator stands, is not in the centre of the circular room, but about half way between the centre and the circumference.[28]

The interventions of anatomical science in this setting were both active (the dissection of the body by the demonstrator) and descriptive, categories that should not be seen as distinct and were rarely uncoupled. William Hunter described the elaborate construction of anatomy theaters to promote hearing and seeing, because the narration of dissection was crucial to situating its display. Charles Bell, some thirty years later, would say of his dissection course in the same theater that "regular and full Demonstrations of the Parts dissected are given; where the Application of Anatomy to Surgery is explained, and the Methods of *operating* shown on the Dead Body."[29] Things were demonstrated and shown at the same time that they were explained; neither the whole body displayed nor the words to describe it stood alone.

Dissection demonstrations, according to Hunter, functioned within medical education by allowing the student to "see the preparatory dissection for every lecture; which will make the lecture itself much more intelligible, and fix it deeper in the mind; he will see all the *principal parts* dissected and demonstrated over and over again [on] a number of bodies dissected in succession."[30] However, even with an abundant source of corpses for dissection, as Hunter clearly assumed when he referred to "a number of bodies," cadavers, too, were only one part of a *system*, not the optimal objects of study that were only approximated through other sorts of display when cadavers themselves were scarce. The complicated mess of a decaying corpse, dissected by a demonstrator, was good for seeing some things—for a sense of scale and overall relationship of the parts—but it was less good for others. Even as a representative of a whole body, the cadaver was complemented by other kinds of display.

The classroom was filled with forms of display that were meant to work together as a sort of interrelated whole, some for training students to see small parts and others to show relationships. At the other end of a spectrum of displays from overly complicated integral bodies sat a type of display that has left little evidence for the historian: the rough schematics and chalk drawings of the classroom. These also required explanation and expert narration, but visually they simplified the complexity of a body and diagrammatically showed its functions. They too helped to explain gross anatomy and relationships between parts. And they were created by the anatomist (figures 5 and 6).

We have little access to these drawings, often made on a board with chalk and erased at the end of a lecture, but they sometimes appeared in issues of the *London Medical Gazette*. It is clear that they were an integral part of the lectures. Bell's wife wrote after his death: "By constant practice he became an attractive lecturer. . . . I have been told that his rapid and effective sketches on the black-board were a great aid."[31] And Bell often wrote in his letters to his brother about making drawings for his class.[32] An account printed in the *London Medical Gazette* of Bell's "Lectures on the Nervous Sys-

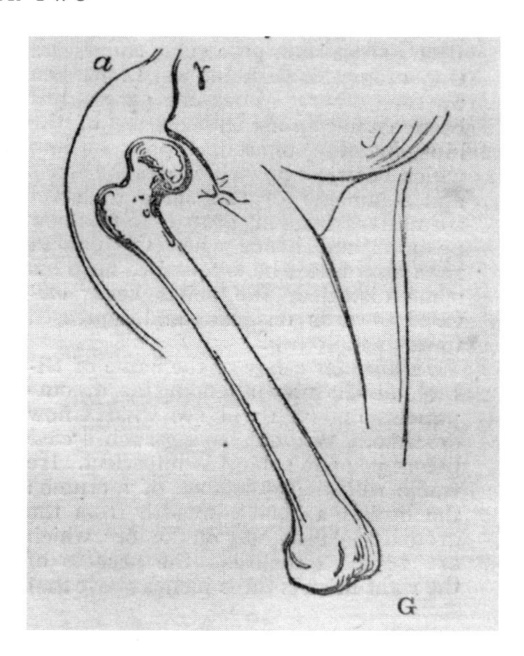

FIGURE 5. A rough schematic of a femur, from Charles Bell, "On the Diseases and Accidents to Which the Hip-Joint Is Liable," *London Medical Gazette* 1, no. 6 (1828): 137. An ephemeral product of the classroom, it represents an anatomist's idea. It was clearly significant enough within the lecture itself to warrant a place in its subsequent publication. Photograph provided by the College of Physicians of Philadelphia.

tem" before the Royal College of Surgeons observes, for example, "Here the Professor pointed to a sketch of the nerves in the Scarabaeus Nasicornis."[33] Another tells us:

> Mr. Bell here referred to his drawings which consisted, first, of the nervous system of the leech, showing a double line of nerves passing in the length of the animal, with ganglions at regular intervals. The second drawing was a view of the sympathetic system in man, showing a series of ganglions bearing resemblance to the former. And in the third was shown the spinal marrow, with all its nerves, beginning above with the fifth pair, and terminating below in the last sacral nerve; these nerves being all double at their roots, and having a large ganglion upon the posterior one.[34]

The notebooks of Bell's students are full of drawings in the midst of class notes (figure 6).[35] These drawings were useful for large-scale relationships and comparisons—whether of the angle of the hip to the femur in a joint or of the nervous systems of leech and man. In the comparisons between two sketches—two ideas about structure—anatomists and medical men could identify anatomical functions and, by analogy and induction, corresponding laws of nature.

It is apparent that the ephemeral classroom drawings that were very much a part of lectures helped to convey the ideas of their maker, what John Bell termed "plans."[36] Charles Bell's lecture associated with the drawing above was on causes of repeated dislocation, and the image appears to depict the angle of the dislocated femur.[37] It conveys a relationship between parts, a rough approximation, an idea. Similarly, the descriptions of Bell's drawings of the nervous systems of leeches and men suggest that Bell was after relationships—something that would be supported by comparative visual illustrations of a rough and approximate nature because they were diagrams or models, not meant to be the objects of study themselves. Finer structures would require other visual tools.

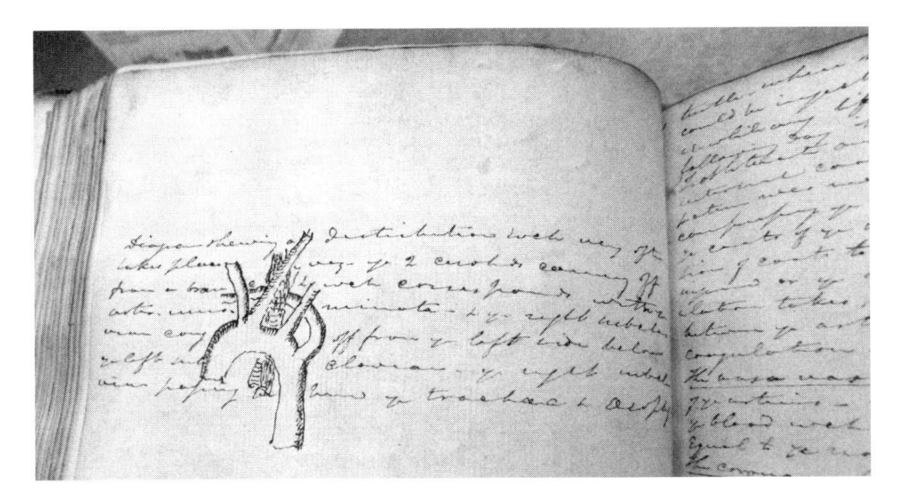

FIGURE 6. Sketch of the arteries found in student notes from Charles Bell's 1822 lectures. This drawing of the arteries, taken from Charles Bell's own sketch, in the notebook of one of Bell's students amidst his textual notes, demonstrates the value of a quick sketch to the students. This student's one hundred ten pages of notes from Bell's 1822 anatomy class contain drawings throughout—sometimes naturalistic and sometimes schematic. Anonymous, "Anonymous Fair-Copy Notes from Lectures on Anatomy and Surgery Reputedly Delivered by Sir Charles Bell in 1822," Leeds University Library, Brotherton Special Collections, MS 595. Reproduced with permission from the Leeds University Library.

SEEING HIDDEN NATURE, HOLDING NATURE STILL

William Hunter had spoken about another element of the system of display used to train the eye when he described passing around the classroom the specimens and preparations selected to depict the finer structures of anatomy. He offered detailed instructions to students, saying: "the *preparations,* must be sent round the company; that every student may examine them in his own hand. . . . [P]reparations are to go round from right to left; in the second bench, from left to right; and so alternately, to the farthest seat of all. To prevent loss of time, when you give a preparation to your neighbour, be so good as to point out the *part,* or *circumstance* which is then to be examined; as I shall do, when it is first handed round."[38] Dissection demonstrations were good for showing students large parts, principal parts, of the body and for showing relationships within the body as a whole, but Hunter himself recognized dissection as functioning directly alongside anatomical preparations that were used in the classroom. His brother John Hunter had amassed one of England's great and systematic collections of such preparations, which had become a necessary part of the anatomy teacher's classroom.[39] Such objects revealed the minute parts of anatomy.

In addition to recognizing their value in the classroom, Hunter discussed the most recent progress in anatomy as being located in these sorts of teaching displays that had been crafted as instructional tools. "Were the great Harvey to rise from his grave, to examine what has been done since his time, I imagine that nothing would give him more pleasure, than to view with attention, the cabinets of some of the Anatomists of the present times. . . . In the latter part of the last century Anatomy made two great steps, by the invention of injections, and the method of making what we commonly call preparations."[40] Preservations, injections, models, and desiccated specimens were all used as objects of study; as a mode of training; as a way of "seeing" systems of barely visible anatomical parts with clarity, away from the messiness of the body; and as a means of maintaining parts for teaching and for research that otherwise too quickly decayed.

Two kinds of object that sat side by side in the medical classroom, fulfilling the same purpose and occupying the same place in the system of display, were wax models and jarred organs (figures 7 and 8).[41] These made the invisible visible and preserved that which was rarely seen in order to train students through visual experience. Hunter described the ways in which they made nature's delicately tiny contents visible in a predictable and consistent way, one that did not depend on significant labor and more significant luck. They were to be placed alongside corpses as displays to be used in conjunction with them:

> Besides dead bodies, we said, that a professor of Anatomy should have a competent stock of *preparations*. . . . Preparations serve two purposes chiefly, to wit, the preservation of uncommon things, and the preservation of such things as required considerable labour to anatomize them, so as to shew their structure distinctly. Of the first sort are, the pregnant uterus, diseases, parts of singular conformation, &c. Of the second class are, preparations of the ear, the eye, and, in general, such as shew the very fine and delicate parts of the body, which we call the minutiae of Anatomy.[42]

John Hunter, John Bell, and Charles Bell, like William Hunter, all kept museums full of large collections of such injected, dried, or jarred specimens,[43] and specimens themselves were greatly responsible for an anatomy teacher's ability to attract students.[44] Simon Chaplin has argued compellingly that for John Hunter such specimens functioned alongside dissection as didactic tools, but that the specimens themselves, in the context of the museum, also helped to construct notions of anatomical displays as a "form of 'natural' spectacle," simultaneously naturalizing the art of dissection as a way of learning about living bodies and conferring "upon dissection a degree of epistemological legitimacy (as a valid way of knowing about living things) in the eyes of 'expert' spectators drawn from [John] Hunter's medical and scientific peers."[45] The specimens were the tools both of students and of experts. They allowed students to see what would otherwise have been invisible in a crowded lecture theater, the instructor demonstrating anatomy through dissection at the front of the room. They captured the possible variations of normal and pathological in a systematic way. And by making them, advanced students contributed materially to classroom research as well as learning mechanical and visual skills.

In his *Two Introductory Lectures,* William Hunter moves from his description of injected vessels to casts and wax models, sometimes also called "specimens" or "preparations" by anatomists in this period. Hunter said of casts, "The proper, or principal use of this art, is, to preserve a very perfect likeness of such subjects as we but seldom can meet with, or cannot well preserve in a natural state; a subject in pregnancy, for example."[46] Casts functioned, in other words, much like actual specimens. The point of these three-dimensional displays was not to act as representations of nature, but to freeze nature herself so that she could be studied. Wax casts were meant to be a "perfect likeness" of the real thing. In the continuum of displays,

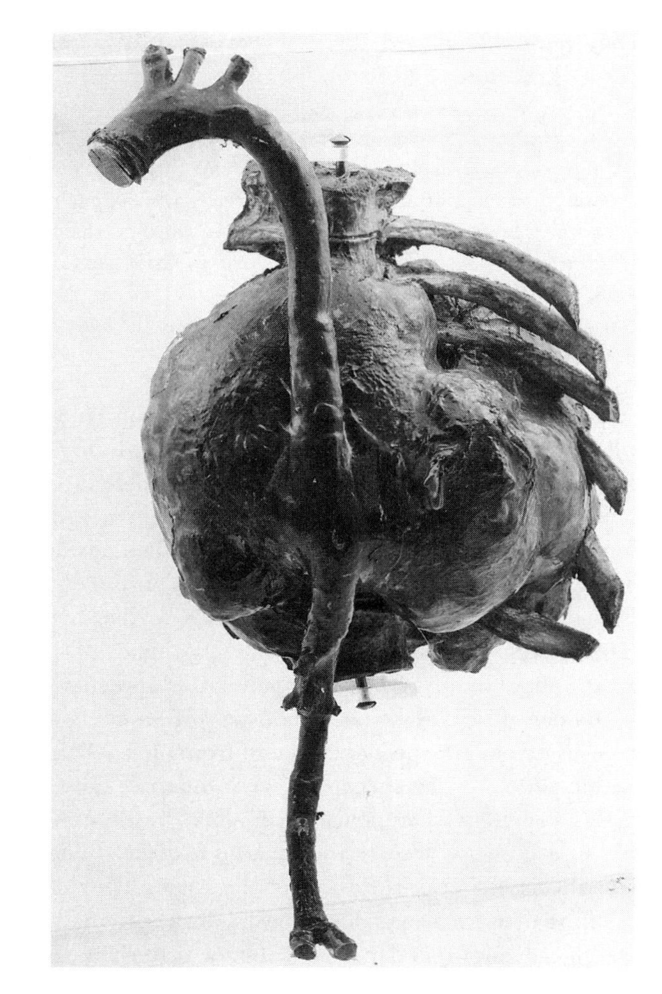

FIGURE 7. Dry specimen of thoracic aorta with aneurysm. This pathological specimen is "dry," though such anatomical specimens were often "wet" and kept in a preservative spirit in jars. A variety of preservation techniques attempted to preserve a lifelike quality as well as the structure of the organ. Royal College of Surgeons of Edinburgh, Bell Collection, BC.xii.2.M.57, GC 11006. Reprinted with permission from the Royal College of Surgeons of Edinburgh.

they served as an intermediate between nature and representation. Though they were wholly crafted by men of science, they were not themselves understood to act as creations or representations. The intervention of science, the labor of anatomists and their craftsmen, was not meant to make an argument or to idealize or to reveal a hidden truth: that intervention was merely made to prevent decay and to hold nature still. In so doing, it rendered accessible to the anatomy student those things that were not everyday, those that were not the normal or typical or ideal. Hunter said of preparations in general, "the object is ready to be seen at any time. And, in the same manner [the anatomist] can preserve anatomical curiosities, or rarities of every kind; such as, parts that are uncommonly formed; parts that are diseased; the parts of the pregnant uterus and its contents. . . ."[47] Even the names "preservation" and "specimen" indicate that these displays themselves—models and specimens—were the object of study: these objects were taken *to be nature* for purposes of instruction in anatomical observation.

This view was not peculiar to Hunter. As Charles Bell's wife recalled, Bell "had discovered a method of modelling morbid appearances in wax retaining their colour in its original freshness, so as to perpetuate for the student much that was lost to them in the usual manner of preserving them."[48] "Colour" and "freshness" were valued in Bell's models because they were things lost quickly in dissections. Again, they offered the opportunity to prevent decay. These man-made wax models were preserving *nature* for the student's observation.

Clear evidence of this can be found in Bell's accompanying museum catalogue. Of the wax model in figure 8, he notes: "From an adult male who survived the operation of herniotomy during several days but without alleviation of symptoms. . . . Though successfully reduced by operation the strangulated loop of intestine was black and gangrenous."[49] Casts and models were made from specific bodies, taken again as ways of preserving such bodies. If jarred specimens were for the body's minute parts, casts and models situated those parts or preserved the not-often-seen anatomies of a slightly larger scale. One could not "jar" a torso or desiccate it, so instead, one modeled it. William Hunter made casts of each of the subjects contained in his atlas on the gravid uterus and displayed them alongside the books, a three-dimensional object preserving nature.[50] Preservations and models were each the products of a great deal of scientific labor—sometimes not even intervention, but creation. But when made successfully, that labor was obscured, such that both preserved specimens and models, or casts, could appear to be direct products of corpses, or sometimes even of living bodies, with lifelike color—the natural objects of study for anatomists.

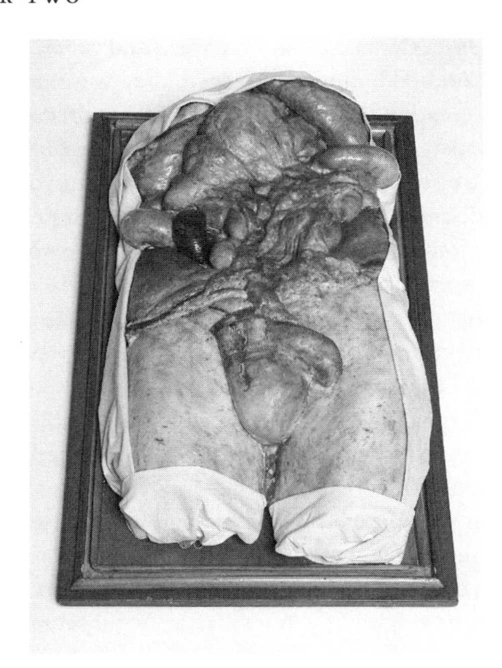

FIGURE 8. Wax and plaster cast of torso. Such casts, taken from individual bodies, provided common ways of preserving the anatomy of larger parts and their relations. This one was taken "[f]rom an adult male who survived the operation of herniotomy during several days but without alleviation of symptoms." Royal College of Surgeons of Edinburgh, Bell Collection, GC 1.43.04. Reprinted with permission from the Royal College of Surgeons of Edinburgh.

The audience for these anatomical objects was wide and varied, as the specimens were often housed in museums that were open to the public.[51] But their primary *intended* audience was medical in nature.[52] They were teaching tools, at a time when objects were, according to both Bell and Hunter, the best ways of "making an impression on the brain."

These three-dimensional objects, thought to be a sort of nature-held-still, must be understood as functioning within the classroom as collections and not as individual pieces. The collections, combining both specimens of the same body part in series to show variation and of different body parts to represent the total organism, display varieties of normal and pathological tissue preserved by various means. Thus, in his *System of Operative Surgery* (1807), Bell wrote: "The cast of this subject and the dissected bladder completes the series of preparations of fistula in perineo, to be seen in my Collection."[53] The preparation was a part of a system of bladders with similar pathologies, as well as normal varieties. Bell said of his museum housing these collections, "It is a room admired for its proportions of great size, with a handsome gallery running round; the class room [*sic*] door opens from the gallery. It would require a month to go round the museum with a book in your hand. I knew that this was a thing to me above all value, and already, by good arrangement, and by the addition of my own preparations, I have filled the room."[54] Bell's museum was connected directly to the classroom, so that classes could take place in its midst and so that students themselves could walk through the museum and sketch the specimens. Some of the specimens were inherited from previous occupants of the Great Windmill Street School and others were of Bell's own creation.

The systematic nature of display that made these objects useful for students included not only the arrangement of specimens together but also a *textual*, as well as a *visual*, context. Scanty but significant text within the catalogues that almost always accompanied museums situated individual objects, and designated a particular way of seeing for an audience of medical students. A general audience could admire museum displays or atlas images in a casual fashion, but these objects were used as tools for examining nature by the audience of anatomists and medical men whose attention addressed the system of display as a whole and perceived the interrelation of its parts. Bell's description of the wax model as having been taken "from" a patient offers one such example of a catalogue entry. In another, describing a dried specimen of a "thoracic aorta with aneurysm," now held at the Royal College of Surgeons of Edinburgh, Bell wrote of the patient's symptoms before death that he "lay long in the Middlesex Hospital being kept very low, and occasionally bled, his sufferings were by no means so acute, as we would imagine must necessarily result from such extensive disease." Concerning the situation of the diseased organ within the individual's body upon dissection, Bell further noted: "Tumour has burst through to the back part, where it formed a very large Tumour during life, notwithstanding the distance of this posterior sac, from the Heart, the pulsation of the Tumour was at all Times very distinct . . . he died exhausted from weakness."[55] The catalogue, an integral part of the display itself, offered a *context* for an individual pathological specimen, detailing what in it was normal and what pathological. Most museums would have had a similar sort of contextualizing catalogue, filled with the history of the specimens being displayed.[56] Captions attached to atlas images played a similar role within the system of display, binding together and situating its elements so that a student could use it to begin to see comparisons.

Finally, the lecturer knitted all the pieces together. Bell was, by all accounts, a good lecturer, and he worked hard at it, spending long hours writing and rehearsing. His wife recalled, "His notes were on strips of paper, with pen scratchings of figures rudely drawn to remind him of the 'heads.'"[57] While dissected bodies and ephemeral chalk drawings helped students to see the relationships between anatomical parts, to see bodies as whole organisms, and museum objects and wax models taught students how to see the minutiae that formed anatomical parts, the lecturer himself taught the student by integrating the elements of the system.

Books and paintings sat alongside anatomized cadavers and bottled specimens in the classrooms of their authors as a part of a comprehensive system of display for teaching medical and surgical students how to see bodies. As Bell described his museum, which featured two-dimensional displays alongside preparations: "My little collection begins now to look well . . . paintings placed in the interstices of the preparations."[58] William Hunter, too, had framed the drawings of his atlas in the context of other visual displays on which he was working:

> The first ten plates are represented in the museum by a number of plaster of Paris casts. These were taken actually from the same subject, and show the same stages of the dissection as certain of the drawings; they were subsequently coloured after nature. . . . The whole of them are exactly nature herself, and almost as good as the fresh subject. We have a good many of them to help us on; they are most useful, especially where it is so difficult to get a subject of this kind to explain upon in a course of lectures.[59]

When representing nature (and Hunter's casts and drawings were depicting the same subjects because they were, in fact, meant to act *as* nature, as object of study, as useful), the plates were objects for use in instructing students of anatomy. They supplemented three-dimensional displays, highlighting minute details that were necessarily obscured in converting dissections into permanent preparations.

The classrooms of medical men were filled with displays that, when integrated by a lecturer, taught future doctors and surgeons to see, to know, and to remember. Each element of the system of displays, some ephemeral and some valued for their permanence, taught students to see and remember the anatomical system in a slightly different way—at a slightly different level or with a different lens—such that when taken together, they formed a sort of comprehensive three dimensional anatomy of a living person. In addition to seeing, however, they needed to learn the use of their hands.

DISCIPLINING HAND AND EYE

The objects of display that provided the first subjects of anatomy training have already been described, along with the process of visualizing that bound them together. The hand, taken to be an sensory/mechanical organ for learning analogous to the eye, was taught simultaneously, as is made evident in Bell's 1833 Bridgewater Treatise, *The Hand, Its Mechanism and Vital Endowments as Evincing Design*,[60] as well as his earlier manual for artists, the *Essays on the Anatomy and Philosophy of Expression* (1806), and his later *Institutes of Surgery* (1838). Bell strongly advocated training hand and eye together—doing was seeing in anatomy. He taught advanced students by having them dissect and preserve their own specimens, in effect having them conduct their own research. He described one of his pupils, "a German physician, who dissected the nerves with extraordinary perseverance," but cautioned that, if "you contemplate a body that has been thus preserved in spirits for three months, and dissected morning, noon, and night, the tissue of nerves which is displayed appears in inextricable confusion. . . . [T]he whole is apparent confusion."[61]

Here, seeing the anatomical system properly required preserving its parts through the creation of displays under the guidance of a masterful anatomist. It also required comparison—the accruing of multiple specimens—and a mindfulness of the relationship between form and function. Bell now offered his audience the solution to the confusion he had noted:

> But when you dissect a second body, and perhaps a third; and when your curiosity leads you to inquire whether a certain part is supplied with one, two, or three nerves in all the bodies, or whether the same little ganglion lodges in the same recess, and receives the same branches in the first and in the second and the third, and you discover that the nerves correspond exactly in every body,—that there is no such thing as a nerve deviating, or being wanting, unless through the hurry or awkwardness of dissection, you are constrained to believe that the confusion is in our heads, and that there must reign a symmetry and a systematic arrangement in the distribution of the nerves.[62]

Bodies could not stand alone intelligibly. Students learned to see by repeatedly and methodically creating and preserving anatomical objects and by drawing, using their hands. It is not terribly surprising that manual skill was emphasized, given that anatomy was the science most closely related to surgery, and, as the historian Andrew Cunningham has argued, before the nineteenth century anatomy was the active and experimental medical science.[63]

In his 1838 textbook, *Institutes of Surgery*, Bell advocated the construction of teaching displays to teach students the manual dexterity and hand-eye coordination they would need as surgeons. "It is essential," Bell observed, "that he [the student] should practise some mechanical exercise, that he may acquire an accordance between the eye and the hand." Indeed, his own brother John had put him to "drawing, modelling, and etching, with this view." But Bell advised that "perhaps the best exercise of all is the art of anatomical preparation,—a very different matter from that exercise of the scalpel with which students are generally satisfied."[64] One reason, he claimed, that anatomical preparation was superior to dissection for teaching anatomy was that "this art of anatomy . . . conveys the knowledge not only of structure but of pathology; for the hasty examinations of the physicians in the dead-house are comparatively of little value."[65] Anatomical displays, the preserved specimens of the sort he housed in his museum, were doubly rewarding to students: their preparation taught them the manual skills and discipline that they would need in the dissecting room and surgical theater, while creating displays also afforded students the time to study, and know by sight, various pathological tissues, thereby satisfying one aspect of the system of display described in this chapter.

Bell also sought to create an intersection in the classroom between the fine arts and anatomical science because he saw mechanical training, or disciplining of the body, as essential to both. He viewed the eye and the hand as similar organs, writing in his Bridgewater Treatise: "we have to show how much the sense of vision depends on the Hand, and how strict the analogy is between these two organs."[66] The two organs, analogous in structure and function, required similar training, both attaining better functioning with age and practice: "in truth, the motions of the eye are made perfect, like those of the hand, by slow degrees. In both organs there is a compound operation:—the impression on the nerve of sense is accompanied with an effort of the will, to accommodate the muscular action to it."[67] Three decades earlier, Bell had devoted much of his *Essays on the Anatomy of Expression* to an assessment of the eye, just as his Bridgewater Treatise would be devoted to the workings of the hand. Both books were written for general audiences, and their basic assumptions—that the hand and eye are analogous and can be trained analogously—carry over to Bell's discussions about surgical training and training in art, as well as to his classroom practices.

A classroom full of anatomical objects and a set of instructional practices that involved viewing them, then describing them through words and sketches, and finally making them, fitted the philosophy of learning presented in *The Hand*. Bell wrote about the process of learning that "[i]n the early stages of life, before our minds have the full power of comprehension, the objects around us serve but to excite and exercise the outward senses."[68] That phase of learning would be the one that involved the casual sorts of interactions with dead bodies, sick and living bodies, and anatomy museums and cabinets that would have constituted the backdrop of experience for a beginning medical student. Then, according to Bell, we train the senses through a conditioning of the sensory apparatus and of the will to exert muscular action, developing simultaneously a linguistic and pictorial capacity to represent those experiences:

> When treating of the senses, and showing how one organ profits by the exercise of the other, and how each is indebted to that of touch, I was led to observe that the sensibility of the skin is the most dependent of all on the exercise of another quality. Without a sense of muscular action or a consciousness of the degree of effort made, the proper sense of touch could hardly be an inlet to knowledge at all. I have now to show that the motion of the hand and fingers, and the sense or consciousness of the action of the muscles in producing these motions, must be combined with the sense of touch, properly so called, before we can ascribe to this sense the influence which it possesses over the other organs.[69]

This was what happened in the medical classroom: first the eye was trained to see; then the hand was trained to know. Students watched dissections and watched their teacher sketch rough diagrams; they looked at atlases; they passed specimens around the classroom; and then they dissected bodies, sketched specimens, and created their own specimens. They ordered the world using hand and eye by creating something with their hands that asserted a visual understanding—asserted that they had learned to see—as well as a manual competence. Their hands had learned and had come to know anatomy. Once the will and the body were trained, then reason and philosophy could be developed through teaching. In that way, learning and the development of science were rooted in the classroom.

It is not surprising, given Bell's views on the relationships between the hand and the eye and between the fine arts and surgery, that he valued teaching anatomy in training artists as much as he did the arts of preservation and drawing in teaching anatomists and medical men. In *Essays on the Anatomy and Philosophy of Expression*, Bell wrote: "The academies of Europe, instituted for the improvement of painting, stop short of the science of anatomy, which is so well suited to enlarge the mind, and to train the eye for observing the forms of nature":[70] as art trained the hands of surgeons, anatomy could train the eyes of artists. It was apparently a controversial point—at the time, the Royal Academy of Arts taught anatomy by having students draw from other drawings and models rather than from cadavers.[71] Objections were probably both practical (bodies were scarce and often required illegal procurement) and philosophical: some thought that artists who were to depict the living should learn to draw from the living or that artistic vision was better fostered by something other than detailed copying.[72]

Bell explicitly confronted these objections to the teaching of anatomy to painters. In *Essays on the Anatomy and Philosophy of Expression,* he wrote: "The study of anatomy has been objected to by some persons of pure taste, from the belief that it leads to the representation of the lineaments of death more than of life, or to monstrous exaggerations of the forms."[73] As a result of such views, the convention at the Royal Academy, much to Bell's chagrin, and that of some of his subsequent students, was to teach anatomy through the study of casts of classical Roman sculpture and of models who posed for the classes. Bell's pedagogical philosophy, which was more than a set of accumulated pragmatic medical teaching practices at work, dictated that artists work with corpses, the natural objects themselves, so as to accumulate manual, sensory experience and learn to see nature, and then to describe her. Bell's student Benjamin Robert Haydon[74] established London's first private art school in 1815 to promote Bell's approach, emphasizing dissection for artists.[75] Haydon had learned from Bell the importance of both human and comparative anatomy, and described his own students spending weeks "hanging over a putrid carcass."[76] He adopted Bell's pedagogical program by teaching detailed, practical anatomy to art students, in Bell's own classrooms when possible, but when that was not possible, with the guidance of Bell's textbooks on dissection. In that way, knowledge could be generalized and ordered in books and texts—objects that were portable and could help to convey a science outside the classroom.

EXTENDING THE CLASSROOM THROUGH BOOKS

While Bell insisted on the significance of classroom learning, illustrated books provided a way of extending the object lessons of the classroom. They fulfilled a couple of different roles within a system of display, depending on the style of book and the style of illustration—both of which determined the price, and therefore use, of the book. Books, through their various uses, allowed Bell to imagine a method of instruction that had an audience extending further than the students in the room. Books made knowledge portable and thus, while they could not replace the object-based learning of the classroom, served as references for physicians and surgeons and exemplified those elements of the visual system that could be embodied in drawings. Portable books were inexpensive and small, illustrated in ways that paralleled the role of chalkboard illustrations, or of an array of jarred specimens. They were good at showing minute structures and their relationships, but they did not stand in for nature. Large atlases could do more, incorporating both the anatomist's sense of how body parts were organized into systems and also including life-size drawings meant to be the anatomists' subject of study itself. Thus, where small books extended the classroom to those at a distance whose knowledge needed updating, large atlases brought together knowledge production and teaching.

Both etching and engraving could be used to reproduce images. Etching was a cheaper technique, which Charles Bell used frequently in books that were designed to be affordable for students and practicing medical men. Those books were meant to be used in conjunction with dissection and other forms of display and might more accurately be termed "reference books" rather than textbooks, as they were not meant to stand alone and were often designed for, and used by, those who had taken on only what irregular training they could afford and continued to try to learn after having begun to practice. John Bell said of texts designed for the student: "when drawings are made for his use, the body should be laid out, as he is to order it in dissection."[77] Dissections were ordered based on how quickly parts and systems putrefied, so texts that were designed to follow this order were designed to do so purely to match the exigencies of dissection and not for reasons of anatomical logic. Laying out a book in that way would imply that it stood as an accompaniment to dissection, at least at some point, and not alone. The book seemed to bring together in two dimensions those elements gross and minute covered by the two separate three-dimensional forms of display, dissections and specimens, taking both as the books' objects.

If books sometimes served as ready references for practitioners, conveying or restating the latest in classroom-based knowledge for an audience that had already acquired experience with anatomical objects or was currently doing so, they also sought to reproduce the object lessons of an anatomy class for students. Sometimes, as in the case of grand folios, drawn to the scale of the human body, the books even acted as nature herself for practicing anatomists. They did so by embracing a warts-and-all style, committing to representing as faithfully as possible the particulars of an individual body as they saw them.[78]

Charles Bell, in a collaboration with his brother John in 1801, observed: "Of [any] twenty bodies not one will be found fit for drawing; but still I conceive that we are not to work out a drawing by piecing and adding from notes and preparations; we are to select carefully from a variety of bodies, that [one body] which gives largeness of parts, where the varieties of parts are well marked, and where there is the most natural distribution of vessels."[79] Bell's remarks express an unwavering commitment to depicting nature exactly as she was found—to selecting well, but drawing the individual body in front of him—and required that he not create some sort of anatomical composite of the "ideal" or "normal."[80] When seeking a body to draw, he looked, as an anatomist would, for "normal" distribution of the parts, but he also kept in mind the requirements of the artist and looked for a body in which the anatomical parts that he was drawing were "well marked" and large. To those who would do otherwise or who objected to the peculiarity of individual bodies, Bell offered text as an antidote, saying, "let us allow ourselves no license but copy accurately. By noting in the description any little deviation every necessary end is answered."[81] Thus the text served to provide indications of what could be universalized.

John Bell, like his brother, adopted the practice of depicting individual corpses and using text to situate them. Describing the plate from figure 9, which depicts muscles of the face, neck, throat, shoulder, and breast, he said, "It was drawn from a subject that had been hanged, and the neck being broken, the head lies flatter upon one shoulder, than it should do even in the dead body, for the Atlas and Dentatus, the two first Vertebrae of the Neck, were fairly broken loose from each other.—The Muscles are more distinctly seen on the left side, on the right side they are thrown into shadow, and are but faintly indicated."[82] The caption clearly demonstrates that text, even minimal text, was hardly incidental. It allowed for the depiction of the very particular—a corpse made the subject of dissection by hanging—in order to convey general knowledge about the normal anatomy of the human body, describing ways in which a hanging victim's neck muscles would be like and unlike those of individuals that a surgeon-anatomist would be likely to encounter in practice.

William Hunter also espoused representation of what was seen directly—a depiction of the natural that preserved the immediate sensory experience and observation of an individual body by the anatomist—and instructed his artists to copy directly from dissected corpses, warts and all. He famously engaged in a debate with the artist Joshua Reynolds about

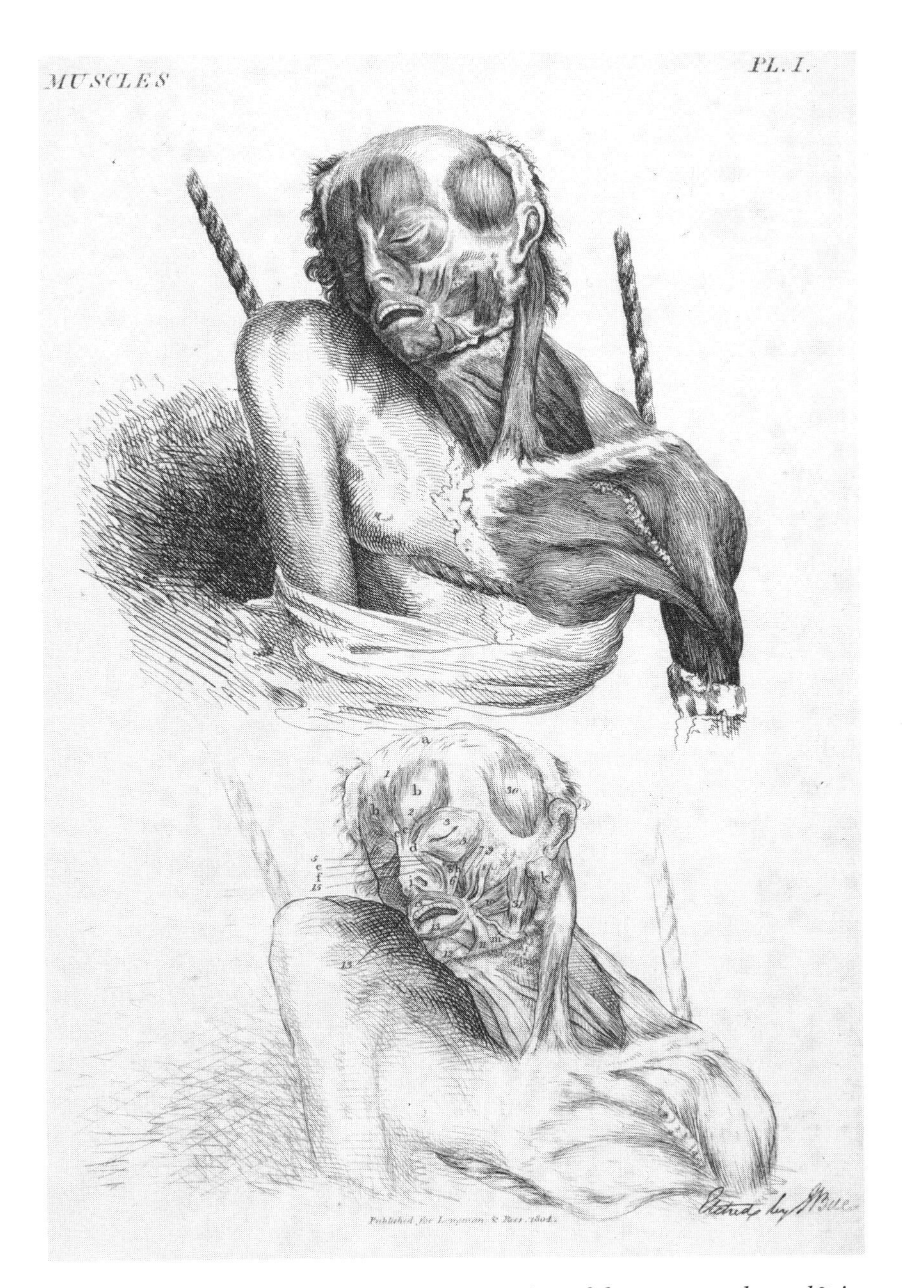

FIGURE 9. Engraving by John Bell in his *Engravings of the Bones, Muscles, and Joints* (London: Longman and Rees, and Cadell and Davies, 1804), book 2, plate II, p. 93. John Bell wrote of this image: "This plate belongs chiefly to the Throat. . . . This Plate explains first all the individual parts one by one, and then joins them, showing how the whole is composed, without which regular form of demonstration, nothing could be clearly understood of parts so very intricate and difficult, and having so long a catalogue of hard names connected with them." Reprinted with permission from the Royal Academy of Arts, London.

whether the copying of nature could itself constitute art, or whether mere imitation without either embellishment or essentializing constituted only a craft.[83] Hunter's verdict, after responding to Reynolds by considering the merits of both the artist's tendency to idealize or universalize and the anatomist's desire for strict accuracy in representing the individual, was that "[t]he one [a faithful copy of nature] may have the elegance and harmony of the natural object; the other [an artist's synthetic rendering] has commonly the hardness of a geometrical diagram: the one shews the object, or gives perception; the other only describes or gives an idea of it. A very essential advantage of the first is, that it represents what was actually seen, it carries the mark of truth, and becomes almost as infallible as the object itself."[84] By representing the object as it was seen, it becomes almost as infallible as nature, and also "gives perception," whereas the idealized image gives only a description or idea of an object—the sort of thing that text could provide anyway. Hunter was sometimes in conflict on this point with his own artist, Jan Van Rymsdyk, but said of his illustrated atlases that they were done with "not so much as a joint of a finger having been moved to shew any part more distinctly, or to give a more picturesque effect"[85] (see figure 10). Like Hunter, John Bell described "a continual struggle between the anatomist and the painter; one [the artist] striving for elegance of form, the other [the anatomist] insisting upon accuracy of representation."[86]

The value placed on copying from nature made these illustrated texts a part of a pedagogical program rooted in experience, and did so in a variety of ways. With the right dimensions, means, and expenditures, the illustrations could act as nature herself, providing detailed, life-size copies of parts of dissected corpses and thus providing students learning to see bodies through a system of displays with an additional set of sensory experiences; they also provided a form of nature to be studied by anatomists developing new knowledge. And alongside these images, which were themselves a sort of anatomical object, sat text—descriptive and sometimes generalizing text. Comprehending as many types of display as possible that were available in the classroom (or at least attempting to fulfill their functions), these books were able to provide sensory experience and to discipline the eye. They also helped to discipline the hand, as students would draw or copy from them much as they did in Bell's museum, learning a manual craft and style of descriptive representation (the stage of learning that followed on sensation). But books also added an opportunity to articulate a plan in nature of the sort that Bell argued could even make the anatomy of the hand interesting to students.

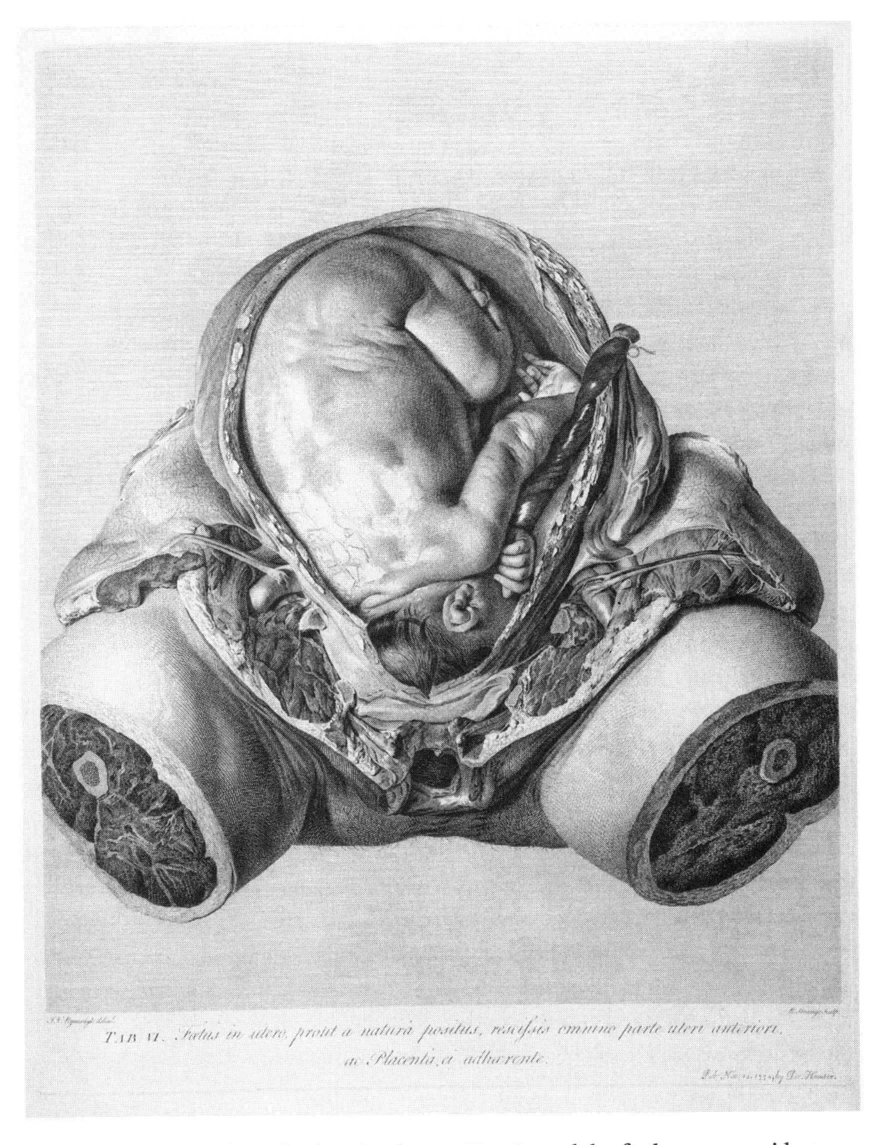

FIGURE 10. Engraving of a drawing by Jan Van Rymsdyk of a human gravid uterus for William Hunter, *Anatomia uteri humani gravidi tabulis illustrata* (Birmingham: John Baskerville, 1774), plate 6, elephant folio. It was about this fetus in particular that Hunter wrote: "... not so much as a joint of a finger [had] been moved to shew any part more distinctly, or to give a more picturesque effect." Image courtesy of the Wellcome Library, London.

UNIFYING BELL'S NATURAL PHILOSOPHY AND ANATOMY THROUGH PEDAGOGY: UNDERSTANDING BELL'S BRIDGEWATER TREATISE ON THE HAND

The system of objects that occupied Bell's classroom, a system that taught students to see and to know, was interrelated and comprehensive with respect to its disciplining of vision and touch. It enacted a pedagogical philosophy that incorporated themes also found in the work of educational philosophers such as the Swiss pedagogue Johann Pestalozzi, whose work was widely known in London;[87] in the practices of private anatomy teachers in London who built museums and anatomy theaters; and in the philosophers of mind of Edinburgh among whom he was schooled and with whom he socialized.[88] Bell combined these different themes in the medical classroom, and they came together as a philosophy in his Bridgewater Treatise on the hand.

In 1829, the Earl of Bridgewater died, leaving behind in his will a sum of money to be made available to the President of the Royal Society of London in order to commission someone to "write, print, and publish, one thousand copies of a work 'On the Power, Wisdom, and Goodness of God, as manifested in the Creation;' illustrating such work by all reasonable arguments, as, for instance, the variety and formation of God's creatures in the animal, vegetable, and mineral kingdoms." Published in 1833, the fourth treatise was Sir Charles Bell's *The Hand: Its Mechanism and Vital Endowments as Evincing Design*. In this pedagogical and natural philosophical Bridgewater Treatise on the hand, Bell relied on the idea that Nature revealed herself to those who studied her, describing in that text a progression of learning that moved from sensing, to naming, to comparing and sorting, to, eventually, understanding purposes and relationships of organisms, and thereby to understanding divine reason. For Bell, who was both deeply religious and, one gets the sense, undogmatic, the sort of natural theology articulated in his treatise provided a guarantee of a fundamental intelligibility built into the world around him. These books became central to the teaching of "safe" (nonradical) science in the years following their publication.[89]

While Charles Bell's Bridgewater Treatise is often taken as a rather ordinary example of natural theology[90] or even of popularization of scientific knowledge, full of the sorts of simplistic explanations that were being replaced by a focus on morphology and evolutionary thought,[91] it should instead be seen as a mature articulation of a pedagogical philosophy. By contextualizing it in that way, we can see *The Hand* as central to the rest of Bell's work and as helping to explain the classroom practices of a man whose room was filled with specimens. Bell's treatise was centrally concerned with the elaboration and application to medical teaching of a pedagogical philosophy in which objects were seen as necessary precursors to books, and knowledge through books was developed first by description of experience and second by arrangement into a regular order of such descriptions. These ideas resonated with various contemporary pedagogical philosophies in Britain.

Bell studied at Edinburgh at a time when moral philosophy had significant standing—when it was taken to be at the core of Edinburgh's Enlightenment. And all his life, Bell sought the opinions and approval of Edinburgh luminaries like Dugald Stewart and John Playfair, sending his brother twenty copies of the one hundred he had printed of his little book on the brain, as discussed in the previous chapter, simply to pass around in Edinburgh.[92] As part of the Edinburgh milieu, Bell inhabited a social and intellectual world in which questions about learning and the ways in which the mind worked were of central concern, and he held in high regard people for whom the answers to such questions involved sensory experience developed into general laws and orders of nature.

It has been said of Stewart that "[h]e offered an image of himself as an explicator of systems, an educator."[93] And according to Stewart himself, at the foundation of a system of natural philosophy, "how far soever we may carry our simplifications, we must ultimately make the appeal to facts for which we have the evidence of our senses."[94] Stewart made reference to another Edinburgh luminary, Adam Smith (evidence of the small and close-knit intellectual community from which Bell came), to explain how the development of knowledge through the senses could be seen in the ways in which language was formed. The first step "would be the assignation of particular names to denote particular objects. . . . Afterwards, as the experience of men enlarged, these names would be gradually applied to other objects resembling the first . . . objects come at last to be classified and referred to their proper genera and species."[95] Bell's pedagogical philosophy draws on similar themes and explanatory mechanisms. Like Stewart, Bell also described the accumulation of sensory experience, the proper linguistic description of sensations and objects, and ultimately the relationships of objects to each other.

Similarly, "object-based learning,"[96] a pedagogical philosophy intended for children and the lower classes, and its proponent, Johann Pestalozzi, had become popular in London as early as 1803, when the *Philosophical Maga-*

zine's "Proceedings of Learned Societies" wrote of the Academy of Sciences at Berlin that there was a paper read "On Pestalozzi's method of teaching, by professor Fischer."[97] A fuller account of his life and teaching methods was printed in the *Athenaeum* in 1807,[98] and Pestalozzi's annual address to members of his school, which appealed for funds for his educational institute, was printed in English in 1818 as "The Address of Pestalozzi to the British public soliciting them to aid by subscriptions his plan of preparing schoolmasters and mistresses for the people, that mankind may in time receive the first principles of intellectual instruction from their Mothers."[99] But what became one of the most widely cited British accounts of Pestalozzi and his methods came from Charles Mayo, whose lecture on Pestalozzi's life and methods, delivered at the Royal Institution in 1826, received wide notice in periodicals.[100] Like Adam Smith before him, Pestalozzi conceived of the formation of knowledge in terms of the accumulation of experience and then generalization and classification. Mayo then described the objects of education in Pestalozzi's program, saying that the first "must be to lead a child to observe with accuracy; the second, to express with correctness the result of his observation. The practice of embodying in language the conceptions we form gives permanence to the impressions."[101] Bell's pedagogical project was clearly one whose themes and explanatory mechanisms resonated with broad philosophical concerns of the day.

Bell had written in his 1806 *Essays on the Anatomy of Expression in Painting*: "It is a fundamental law of our nature that the mind shall have its powers developed through the influence of the body; that the organs of the body shall be the links in the chain of relation between it and the material world, through which the immaterial principle within shall be affected."[102] He explained the virtue of the hand as the subject for a treatise because it was sensory and acted through the force of the will directly upon the material world, allowing its user first to develop proper observational and manual skills, then to understand principles of a plan of nature, and finally, through that development of natural knowledge, to know God. Thus, in this articulation of a pedagogical philosophy, Bell both described the need for object-based instruction in anatomy and also explained how, in consequence, the student generalized such experience to create knowledge.

Bell regarded the hand as a challenge to the anatomy teacher (one could assume all the more so in a text for a general audience)—lots of muscles and little bones to account for in a rather ordinary-seeming appendage: "The demonstration to the anatomical student of the muscles of the human hand and arm, becomes the test of his master's perfection as a teacher.

Watercolor by Charles Bell of a wounded soldier with a missing arm, inscribed "XIII, Waterloo. . . ." This was made from Bell's sketches of wounded soldiers whom he treated in Brussels in June 1815. From the Royal Army Medical Corps Muniment Collection. Image courtesy of the Wellcome Library, London.

"A lecture at the Hunterian Anatomy School, Great Windmill Street, London," watercolor by Robert Blemmel Schnebbelie (1830). One can only imagine that such a scene was inspired by the Great Windmill Street Theatre during Bell's London years. Note the various objects and drawings arranged together in the classroom—a bust in the middle of the room, bones mounted on the chalkboard, a wet preparation on the lecturer's table, the skeleton suspended, and drawings all around. Image courtesy of the Wellcome Library, London.

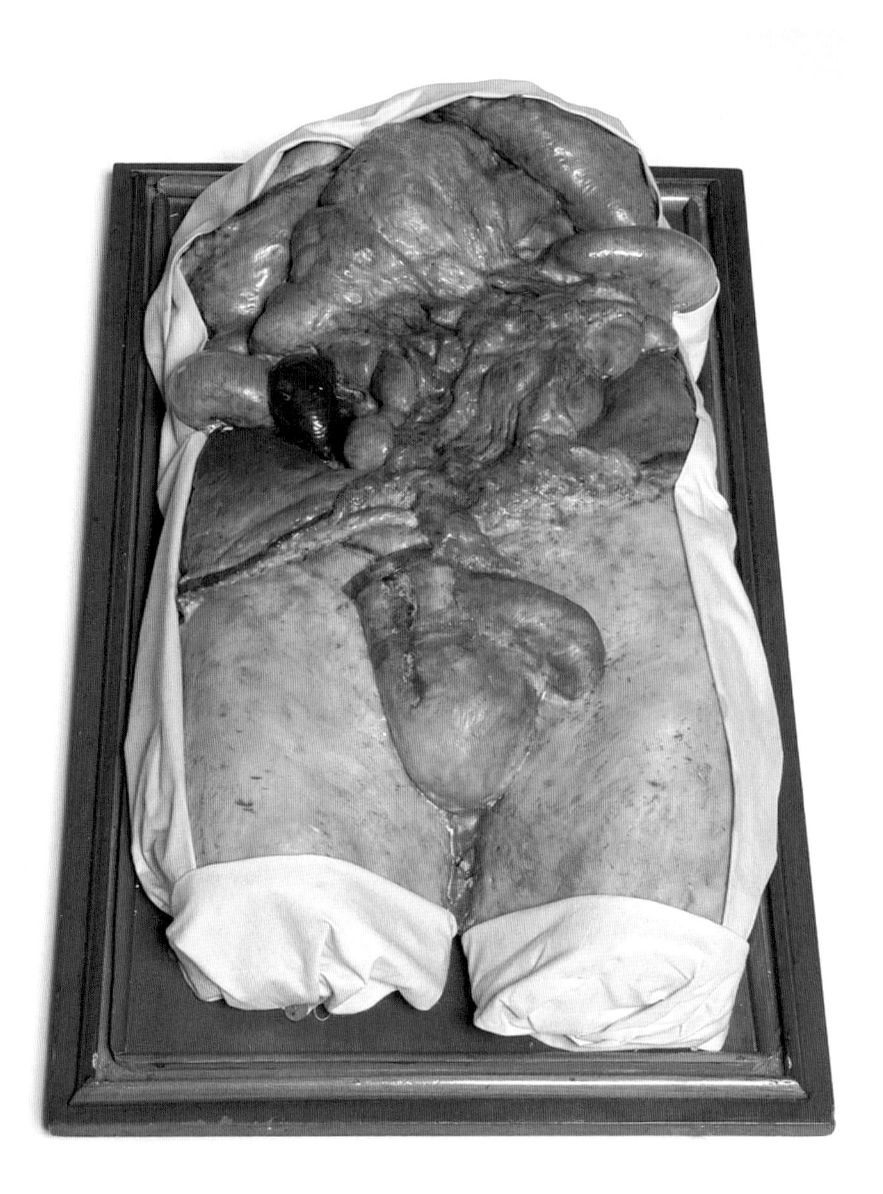

Wax and plaster cast of torso. Such casts, taken from individual bodies, provided common ways of preserving the anatomy of larger parts and their relations. This one was taken "[f]rom an adult male who survived the operation of herniotomy during several days but without alleviation of symptoms." Royal College of Surgeons of Edinburgh, Bell Collection, GC 1.43.04. Reprinted with permission from the Royal College of Surgeons of Edinburgh.

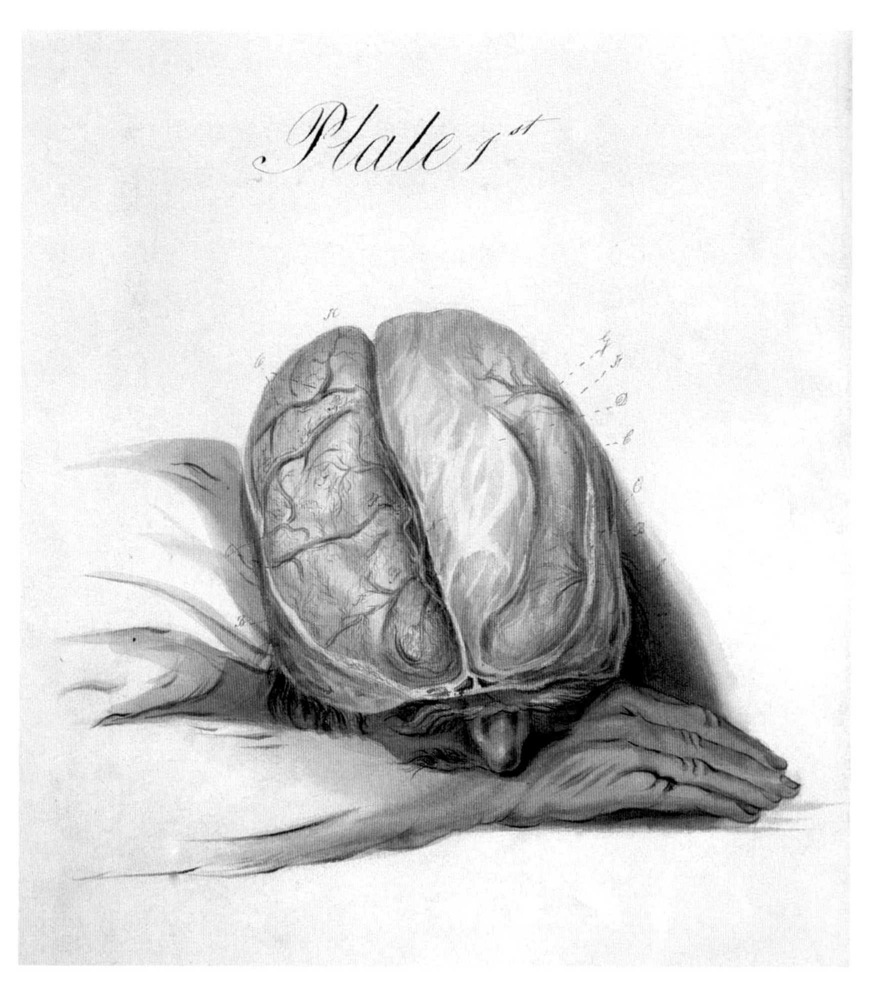

General anatomy and subdivisions of the brain, drawn with the skull cap removed, originally drawn by Charles Bell for Charles Bell, *Anatomy of the Brain* (London: Longman, Hurst, Rees, and Orme, 1802), plate 1. This 1823 manuscript version is hand colored after the original drawings. Image courtesy of the Wellcome Library, London.

Nothing is more uninteresting, tedious, and difficult to attend to, than the demonstration of the muscles of the arm, when they are taken successively, as they present themselves."[103] The good teacher's secret, it seems, was to take these muscles as a part of a broader arrangement, an arrangement with an intelligible order that coupled form and function. In Bell's own words, "when they are taught with lucid arrangement, according to the motions performed by them, it is positively agreeable to find how much interest may be given to the subject."[104]

This sort of inductive reasoning and methodical arrangement of knowledge also made natural theology, itself potentially a sort of simple, descriptive natural history, into a part of a properly philosophical, pedagogically rooted science. Dugald Stewart had done something similar, explicitly arguing in his third volume of *Elements of the Philosophy of the Human Mind* (1827) that, while the logic one can see in nature only *suggests* a Creator, nonetheless it can be useful for students to assume a Creator and to rely on induction from observation to determine nature's laws—it having been taken as a given that nature would be lawlike.[105] He goes on to say that laws should be seen "merely as general rules or theorems obtained by induction" from observations and that sensory observation offered the only bedrock of sound logic—there was no elementary set of principles from which other laws could be deduced. Such thinking, treating the argument from design as a useful theorem in order to be able to make sense of observed phenomena, is fundamental to Bell's *The Hand.* In that sense, natural theology is not the point of the book. It is an assumption (though a very important one and one that Bell held with strong conviction) that underlies the text, taken as a given and not its end goal, or ultimate purpose.[106]

Natural theology provided the guarantee of intelligibility and one way of legitimizing inferences of function from form; observation and sensory experience (in Bell's case, the experience of objects in the classroom) provided the "forms" to which that philosophy could be applied, and from which that abstraction of laws was derived. The basis of knowledge began with sensory observation of, and experience with, nature. Bell connected the two—the descriptions of objects and the view of an interrelated, structured whole—by saying: "If a man contemplate the common objects around him—if he observe the connection between the qualities of things external and the exercise of his senses, between the senses so excited, and the condition of his mind, he will perceive that he is in the centre of a magnificent system, and that the strictest relation is established between the intellectual capacities and the material world."[107]

The table of contents for *The Hand* reflects this philosophy of learning. The hand is defined ("Chapter II. Definition of the Hand"), it is compared ("Chapter III. Comparative Anatomy of the Hand"), and conclusions are drawn ("Chapter X. The Hand Not the Source of Ingenuity or Contrivance, Nor Consequently of Man's Superiority"). The progression of Bell's argument in *The Hand* follows his own philosophy of learning. In so doing, Bell took the system of display that constituted the teaching practices common to the anatomy classrooms of anatomists—it would seem natural, after all, that the substance of surgical training would involve training the hand and eye using anatomical objects—and theorized them. In *The Hand*, he posited that students generally, like students of the hand's complicated anatomy, learn better when presented with an intelligible plan of nature that can be observed than by engaging in rote learning. And in lectures to a more specialized audience, the Royal College of Surgeons, he wrote of the hands-on dissection of the complex system of the nerves, "Now the desire to find the clue to this labyrinth naturally arises. The origin and distribution of each nerve must surely explain its function and use: therefore the relations of the nerves must be like a language: and how happy should we be to find a key that made the characters of this language intelligible!"[108]

Bell's Bridgewater Treatise elaborated, at the same time that it enacted, the pedagogical philosophy of a British anatomist and medical man who rooted his medical science in education. One of the cornerstones of that philosophy was the accumulation of sensory experience concerning material objects. The hand was sensory and acted through the force of the will directly on the material world, allowing its user first to develop proper observational and manual skills, then to understand principles of nature's order, and finally, through that development of natural knowledge, to know God.

CONCLUSION

In Bell's description of the progress of learning in *The Hand*, he elaborated a pedagogical philosophy in which learning from books paralleled experience with other kinds of objects. The knowledge represented in the text was itself developed first by description of experience and second by arrangement and ordering of such descriptions. Bell thereby connected the two—the descriptions of objects and the view of an interrelated, systematic whole. That methodically developed approach to learning—like Stewart's use of method as providing the scientific basis of a philosophy of mind—defined a comprehensive and unified science of medicine. Such seemingly disparate projects as Bell's Bridgewater Treatise on the hand, *Essays on the Anatomy of Expression in Painting* for artists, lectures before the Royal College of Surgeons, and the building of collections of anatomical specimens, as well as what he considered to be his defining scientific achievement, the discovery of the separate roots of motor and sensory nerves, were bound together by a common pedagogical method. It was a method developed through systems of display that required a proper progression of thought: from the observation of anatomical objects, to their description in words and images, their manipulation by hand, and their comparison side by side. Only then could the anatomist generalize about functions and lawlike behaviors that revealed an underlying, elegant and intelligible, nature fashioned by a divine hand.[109]

Bell wrote of his engravings of the arteries: "By long attention to the subject I hope that I have been able to make these Plates simple, intelligible, and accurate . . . the design of this book of Plates is to present to the student, at one glance, the general distribution of the vessels, and to fix them in his memory in a way which no description can accomplish."[110] Bell here informs his reader that his text was designed for students; that its plates were based on carefully selected subjects; and that Bell hoped that the plates were *simple, intelligible,* and *accurate* in order that they might "fix them in [the student's] memory in a way which no description" could. They were meant to be simple and beautiful so that they would be memorable—so that they could be learned.

Anatomists such as William and John Hunter had relied on a similar system of display, teaching anatomy with the same constellation of objects later used by Bell. But while Hunter's drawings existed in elephant folios and presentation copies, Bell's were made widely available to students through journals and inexpensive books—if images were a part of a pedagogical philosophy, they had to be available to students.

For Bell it was imperative to make of medicine and anatomy a unified science. Such unity could be found in particular pedagogical methods. For Bell, the material of life was rendered knowable through sensory experience. Such experience, carefully ordered in the classroom to reflect the order of nature and the order of learning, was presented in display and described in, and brought to bear on, books like *The Hand,* in ways that highlighted the beauty of creation and left a lasting impression. And impressions on the mind were the foundation of learning.[111]

CHAPTER THREE

From the Anatomy Theater to the Political Theater: Journals and the Making of "British Medicine" in Early Nineteenth-Century London

According to many opinions, Great Britain is behind all the other European states in the organization of medical instruction. English legislation on schools of medicine and surgery, on hospitals, and corporations of physicians, surgeons, and apothecaries, as also in regard to the individual privileges of the members of these professions, is truly in a state of chaos. Charles Bell, however jealous of the dignity of his art, could not deny this; but neither for the public nor for his private interest did he approve of the measures attributed to the Government; because the Government itself, not having any settled plan on the question, felt continually obliged to modify under alternating influences. Charles Bell apprehended, above all, both on his own account and on that of other physicians who had won their rank by long study and practice, the pretensions of inexperienced upstarts in science, and of the municipal corporations who wished to regulate the medical art like any other commercial speculation open to a licence.

AMÉDÉE PICHOT, 1860[1]

By the 1820s, Bell had accrued additional successes in London. His work on military surgery and on the nerves, begun during those first few years in London, continued apace. And in 1815, Bell traveled to the Continent again to treat wounded soldiers, this time after the Battle of Waterloo, where he operated for three days straight, performing perhaps hundreds of amputations.[2] He also continued to develop his work on the nerves, still thinking it would revolutionize the anatomy of that system. "This business of the nerves," he wrote, "will be long of coming forward exactly as it should be, but my own ambition has a rest in this, that I have made a greater discovery than ever was made by any one man in anatomy, and the best of it, I am not done yet. I have just finished my paper on the nerves of the face for the Royal Society."[3] That paper was given in 1821.

But perhaps the most considerable honor of this period was Bell's appointment to the Professorship of Anatomy and Surgery of the Royal College of Surgeons in 1813 (figure 11). Bell was especially pleased to report that previous occupants of the position had been Sir William Blizzard, John Abernethy, James Wilson, and Thomas Chevalier—fine company indeed—and called the professorship "the only distinction the profession can bestow."[4]

Bell described his anxiety about his first lecture before the College, saying, "You stand in a very awkward situation—500 people, stretching upwards, bolt before you, and nothing beside or around you. I found the Theatre crowded to suffocation . . . indeed, he must have more confidence in himself than I possess who could, without some misgiving, address the united profession."[5] But despite a dryness of throat and the feeling that his tongue stuck to the roof of his mouth, he earned "lots of compliments from the old gents,"[6] and his lectures went over so well that his audiences got bigger as the course of lectures went on. Charles wrote to George, "There never was anything like the packing and crowding of the Theatre, and yesterday they say as many went away as got admission, and there they sit from three, when the doors open, till four."[7] It was not a foregone conclusion that audiences would fill the gallery and clap politely: Bell described a later Hunterian Oration by Anthony Carlisle (his former rival for the Royal Academy professorship), saying, "He began with a supercilious confidence, and, after many interruptions, finally broke down, after an hour and a quarter's delivery, amid the noise and hisses of the audience. This was a fearful lesson to a lecturer."[8]

In 1814, Bell was elected to the post of hospital surgeon at the Middlesex Hospital, a position that required Bell to put to good use the networks of patronage he had been cultivating over the past decade. Bell had to canvass for votes for the post and wrote to his brother: "I am told I have six hundred people to see, to visit . . . Dressed to a T; list made out, and cards arranged . . . Everything goes on well. The show of votes increases. I have it three to one, but still I must keep moving, and a most unpleasant business it is calling on people."[9] He won the election 369 to 104.

Professional politics remained very much a part of Bell's life in London. So did financial woes.[10] But Bell's classes were full, with eighty to one hundred twenty students—bigger crowds than he had ever had before—and he found himself "at the height of [his] ambition in regard to teaching."[11] Having become more a part of the London medical scene—a person known through his teaching and also through social calls like those required for the canvass—Bell began to express opinions on those things he knew best. It was during these years that Bell was to become a reformer. His reform efforts were centered on pedagogy (reforming of the London curriculum,

FIGURE 11. Lithograph of Charles Bell lecturing at the Royal College of Surgeons. In 1824, Bell accepted the Professorship of Anatomy and Surgery at the Royal College of Surgeons, and by his own accounts, the most renowned of London's senior surgeons could be found in the audiences who attended his popular lectures. Image courtesy of the Wellcome Library, London.

institutional structures, and credentialing) and on personal career building in medicine.

By the late 1820s, his reform efforts were captured in the new medical weeklies that printed accounts of Bell's lectures, in particular, the *London Medical Gazette*. Those journalistic accounts brought classroom practices to a public stage. The journal's endeavor was one that Bell encouraged, though backing for the publication of lectures in journals was not uniformly shared: "You may perceive in a new weekly, 'Medical Gazette,' Nos. 2 and 4, my clinical lectures taken; tell me how they strike you, and suggest any thing. This is a thing we must support."[12] There, his lectures on organs of the voice, on the hernia, on the larynx and laryngotomy, on diseases of the urethra and bladder, on lithotomy, esophagotomy (the removal of foreign bodies from the esophagus), amputations, disease of the hip joint, and roughly a dozen other topics were published. Teaching became public in new ways in the early nineteenth century, as new weekly journals filled their pages with news from the classroom. Through those journals, pedagogy and politics, intertwined in the kind of private network building Bell participated in to fill his classrooms, were bound together. The published lectures, and the medical journals containing them, sometimes helped bring Bell's pedagogical philosophy to a broader stage, but also constituted the changing backdrop beyond his control against which Bell's career played out.

Medical weeklies attempted to shape political communities and constituencies in 1820s and 1830s London in order to solidify readership, and to some extent they succeeded. The *Lancet* tended to draw more radical reformers, whereas the *London Medical Gazette* was known as a more conservative journal, but those communities were always imperfectly aligned in a world where politics was driven by self-interest, social patronage, and professional and institutional affiliations as much as by ideologies.

PEDAGOGY BY OTHER MEANS: THE BIRTH OF MEDICAL WEEKLIES

When Charles Bell printed his little book *A New Idea of the Anatomy of the Brain* in 1811 and circulated it among his friends, publicizing was not the same thing as publishing. Bell knew his audience and his community, a community developed in shared spaces through direct interactions. Though periodicals circulating to a general audience were available by the beginning of the nineteenth century, the majority of British medical periodicals, in particular the weeklies that developed wide audiences, were not born until the 1820s.[13] When medical periodicals did begin to circulate, they did so widely, with several competing journals becoming available within just a few years of each other. The majority of these medical journals were published in London.[14] Although medical periodicals started being printed later than other kinds of journals, they were a part of a general trend in publishing: the number of periodicals in Britain tripled in the first three decades of the nineteenth century.[15] Thus, by 1830 pedagogical practices had come to include periodicals—teachers recommended that students read particular periodicals,[16] and many assumed that their lectures would appear later in print.

Through these new reading practices, journals brought people into contact in new ways, opening up medical education and bringing the contents of classroom practice itself to a wide reading public. Pedagogical practices came to encompass publishing, revealing assumptions that knowledge could be effectively disseminated through text. While in 1812, when Bell bought his share in the Great Windmill Street School, students learned about various schools through word of mouth or advertisements, by 1830, a medical student in London looking for guidance or for the contents of London's medical lectures could turn to a variety of periodical sources. Taken together, the amount of information to which such students had access was almost overwhelming. Politics helped to distinguish the primary journals that competed for a student's or professional's attention, hoping to create loyalty through a political community that they themselves were trying simultaneously to build.

The two primary medical journals in London, the *Lancet* (generally thought to be radical and founded in 1823) and the *London Medical Gazette* (regarded as conservative and founded in 1827), were both weeklies. Weeklies were the most likely journals to be politicized, as they were published frequently enough to include relevant and timely political commentary as well as to have space for correspondence.[17] Both the *Lancet* and the *London Medical Gazette* produced almost two thousand pages per year and were established with defined political positions. In addition, medical men could turn to the *London Medical and Surgical Journal* (founded in 1828), which espoused radical politics and was published as a monthly until 1832, then becoming a weekly; the *Medical-Chirurgical Review* (founded in 1820), an expensive, moderate quarterly; and, of course, the nonspecialist journals like the *Edinburgh Review*, in which matters of general interest were published.[18] Bell's contemporaries recognized the political nature of journals readily. The editors of the *London Medical and Surgical Journal* wrote about the politics of major medical journals with some disdain in 1834:

> It is generally the policy of a journal to set itself up as the advocate of this or that party; and, by a devoted attachment to its interests, by respecting or lauding its prejudices, by denying or palliating its defects, and, above all, by heaping opprobrium upon the antagonist faction, to earn an interested support. Of the nature of this exaggerated advocacy in state politics, the public is so well aware, that none but the most violent adopt the opinions, or credit the unauthorised statements of a newspaper on either side.[19]

The political terrain and alliances that these journals appeared to create was different from that which had defined the medical community as a social group previously, seeming to constitute a politics shaped by a new, virtual community.[20]

When Charles Bell first arrived in London, he attempted to make a place for himself professionally by asking for letters of introduction to prominent members of the community from well-placed friends back home, as well as by attending dinners and other social functions.[21] This was how one got a foothold in the community and found students and open hospital positions. The new journals, however, brought together social networks defined by professional medical politics in new and different ways that small face-to-face gatherings would not have allowed. New, print-dependent communities often crossed the formerly all-important boundaries of professional rank, education, and family background, bringing together individuals who might not have met in person and helping to align their interests. But while the journals seem outwardly to suggest a set of ideological medical politics that defined communities of practitioners in the period, a closer examination reveals that transitions were slow and that the communities these journals seemed to reflect were actually only just being created by the journals themselves. If journals espoused particular political positions and rhetoric, the groups they served were still made up of individuals moving through the same systems of patronage as their predecessors had—individuals whose politics continued to be flexible, and whose affiliation with a community formed through a journal was also sometimes loose.

REPRODUCING THE CLASSROOM AND CLINIC IN PRINT

One way that the medical journals filled their two thousand pages per year was to print the contents of lectures given around town. Sometimes they printed those contents with the permission of the author, who would even edit the lectures, written up by students or others in attendance. But in other cases, journals, particularly the *Lancet,* would print lectures without the permission of the lecturers. Thomas Wakley famously feuded with John Abernethy, as well as with Astley Cooper, about the publication of lectures that Abernethy and Cooper did not want printed.[22] The lectures would become the central feature in each issue of the journal, with a series of lectures extending across multiple issues of the journal. They would be featured alongside advice to students, clinical reports from the hospitals, reports from provincial or foreign medical societies, opinion pieces by the editor or by letter-writers (sometimes including letters written disputing the contents of the lectures being published), and professional happenings (e.g., lawsuits, deaths, and new teaching positions).

On this issue, as on most others, the *Lancet* and the *London Medical Gazette,* and the authors who contributed to each, espoused different positions. John Abernethy summed up the argument against such publication, in his *Lectures on the Theory and Practice of Surgery,* by laying out the rights of the lecturer:

> If a person educate himself with a view to become a teacher in any department of science, he endeavours to collect, by reading, all the scattered knowledge that has been obtained; to acquire by his own observations and experiments additional information; and to arrange and display the whole of his subject in a perspicuous and impressive manner. Should certain portions of his lectures seem worthy of general attention, he progressively publishes them; and some of the most instructive books in our profession, as they were the result of long-continued meditation and enquiry, have been thus produced. [23]

The unauthorized publication of lectures therefore amounted to theft of the lecturer's property and could only have bad consequences. After all, "who would labour in this manner, under the persuasion that the fruits of his exertions might be surreptitiously taken from him?" Furthermore, if allowed to continue, it would "put a stop to such efforts, and materially impede the progress of science."[24]

Abernethy went on to malign the character of Wakley, calling him a person "so devoid of all good feeling and all sense of shame as to avow and defend conduct so unprincipled."[25] The issue came up again in a trial entirely unrelated to the publishing of medical lectures. In the lawsuit *Bransby Cooper v. Thomas Wakley,* which had to do with accusations against Cooper of malpractice, Cooper used Wakley's publication of other men's lectures to demonstrate that Wakley was of a bad and untrustworthy character, motivated by profit, suggesting that, "for the purpose of committing plunder on the property of others, to assist himself . . . [he did] that which would render it unnecessary for the pupils to attend the lectures, because all the advantages derivable from their attendance might be gained by reading the reports of them in *The Lancet*."[26] The profiting from the work of another, the use of unfinished or unpolished work, the disincentive to pursue scientific advances that were already reported on, and the potential effect of rendering the classroom unnecessary, all were issues that made the publication of lectures without the permission of the lecturer objectionable to those who accused Wakley of damaging the medical community.

Wakley, on the other hand, argued that he was a "friend of a FREE MEDICAL PRESS," and that he was publishing lectures to benefit medical practitioners generally and general practitioners in particular, to allow them to stay abreast of the best and most current medicine and surgery being taught in London. He also argued that the lectures that he published were public anyway. Wakley defended himself at length in an 1830 column entitled "Address to the Readers of *The Lancet*," writing:

> One of the accusations most constantly directed against us was, that we had published without consent—in fact, had stolen and published for our own profit, the lectures of several medical teachers. For five years we treated the accusation with silent contempt; and having thus shown our feeble opponents, that it was not in their power to lessen the influence, or decrease the sale of this work . . . [we choose] to make the profession acquainted with the circumstances under which the whole of the lectures had appeared.[27]

The column reminded readers that there was a distinction between a public teacher and a private one, saying that the *Lancet* recognized the lectures of private lecturers as private property and thus had never printed such lectures without the lecturers' consent. The *Lancet* determined a lecture to be public when it was delivered within a public hospital; lecturers had always charged for attendance at their lectures in hospitals as well as in private buildings, but the site of the lecture being in some way public defined the nature of the lecture for Wakley. He applied this principle in the dispute with Abernethy, where it was upheld in court. Wakley then listed the names of lecturers who had consented to have their lectures published for the sake of "public utility" and in some cases had even assisted the journal by editing the articles.[28] It quickly became clear that any lecture delivered in a public venue was likely to be printed somewhere and it therefore made sense, at least to some, to cooperate and oversee the publication of one's work.

Another practice that was begun by the *Lancet*, and that eventually became ubiquitous, was the publication of hospital reports. If the reprinting of classroom lectures offended some, reproducing clinical experience through print was even more controversial. The *Lancet* began reporting on cases in London's charitable hospitals in November of 1823, and by November 16 of that year they had reported on the first case that ended in a fatality.[29] Cases were narrated as serials, sometimes suspensefully stretching across several issues. Wakley saw his role in printing hospital cases as serving two ends: not only could they serve to educate practitioners as to the nature of contemporary practice in London, but they could also highlight corruption and incompetence in London's hospitals. In an 1899 biography of Wakley, Samuel Sprigge wrote that hospital surgeons "might conceal things awkward to themselves or their hospitals, in which case it would be *his* [Wakley's] duty to reveal them."[30] It did not take long for a variety of objections to surface.

The hospitals realized almost immediately that Wakley's reporting would not serve them well, since Wakley was inclined to highlight the worst elements of hospital practice. Several hospitals quickly barred him, or anyone they discovered who reported to him, from their grounds.[31] The *Medico-Chirurgical Review*, a quarterly edited by James Johnson and one of the few competing medical journals in 1823, wrote of Wakley's practice: "No man can command success in surgical operations; and if a surgeon fails from want of dexterity he suffers mortification enough, heaven knows! in the operating-room, without being put to the cruel and demoniacal torture of seeing the failure blazoned forth to the public!"[32] Wakley mocked Johnson's characterization of the fragile surgeons, but the *Lancet*'s hospital reports continued to be the subject of criticism, for both style and substance, throughout the decade.

The *London Medical Gazette*, which published its own column of hospital reports, wrote a multipart commentary on the *Lancet*'s use of hospital reporting. The article was particularly critical of the unserious style of the *Lancet*'s column: "The purpose of such reports has no characteristic of honesty about it: its object is not to communicate information—because the simplest statements would answer that purpose—but to *attract*; and where one reader attends to a dry record of facts, ten we know will be gained by *embellishings*—especially when these involve the character and conduct of eminent individuals. *Misrepresentation is the main-wheel of the machinery.*"[33]

In addition to the criticism of the *Lancet* for not being properly serious, for being written for the entertainment of audiences rather than for the education of practitioners, the *London Medical Gazette* accused the *Lancet* of improperly making the private public. "We deny," thundered the *Gazette*, "that the treatment of disease is a thing that falls under the cognizance of the public judgment, or ought to be brought under their notice." The public was not competent to evaluate such things: "[O]f medical and surgical matters the public are singularly and pre-eminently ignorant, and of course are singularly and preeminently liable to be deluded. . . . That, again, is not public which is not practised with open doors. That, again, is not public which is accessible only upon payment of fees:—what is *so* attainable, is strictly a private concern."[34] The *London Medical Gazette*, in other words, argued that hospital reports should not be written in a style that might appeal to the public because the knowledge contained therein was only accessible to experts. The argument about what was private and what was public had crucially to do with audience. And again, the implications of extending the classroom through the printing of competing journals with the aim of broad circulation attempted to refashion social groupings and politics.

According to the *London Medical Gazette*, hospital reports had the potential to be sensational and could be presented in an entertaining or scandalous manner, so they had to be carefully crafted in such a way that they would be read only by their appropriate audience—the specialized audience of medical experts who could understand and learn from them. The *London Medical Gazette* also argued that hospitals were not public spaces, that they were not institutions with open doors, and that the lectures the hospitals hosted were also private because students paid to attend them.[35] This argument mirrors the argument made about the reprinting of other medical and surgical lectures. Even early issues of the *London Medical Gazette* were filled with lectures, much like the *Lancet*, but the audience for the journal was conservative in nature—the *London Medical Gazette* seemed simply to want to replicate in print the sorts of professional groups that had been established previously in person, keeping its audience bound by profession, expertise, and rank, unlike the radical *Lancet*, which seemed to challenge all boundaries.

These early articles changed the nature of medical teaching in the middle of Charles Bell's career, just as he had begun to make it within a world defined by the old rules. Journals made it possible to survey the contents of a course without actually paying to attend it, but they also made it possible for a teacher to build a reputation through print rather than through word of mouth, as students could select instructors based on their publication record. Medical weeklies brought the most recent medical developments and lectures of London to practitioners in the provinces, practitioners who had previously been disconnected. It is clear that some lecturers actively courted publication in these journals as a way to publicize their courses and their work,[36] while others saw the journals as threatening classroom attendance or, more commonly, sales of later publications in book form.

Charles Bell, who had been looking for benefactors to pay for his publications since the earliest days of his London career, seems to have embraced medical journals by the mid-1820s, as he was a regular contributor to the *London Medical Gazette*. Of course such printings did not allow for the inclusion of all the displays that constituted so much of classroom experience, but they did attempt faithfully to reproduce the words of the lecturers.

In some ways, therefore, the classroom, at least the public classroom and the classroom of the consenting private lecturer, was opened up, its audience expanded to include even some provincial practitioners. Most often, lectures were printed in the journal whose politics accorded with those of the lecturer, as when a lecturer sought to have his lecture printed, he chose a journal to which he would give access and permission and with which he would align himself. On the one hand, the journals expanded the audience for medical lecturing, bringing together doctors, surgeons, and apothecaries in a forum that united medical professions in a way unlike any that had existed previously; on the other, they also splintered the medical community in new ways, ways that were explicitly based on politics and not on medical profession (apothecary, surgeon, or physician), place of birth, or social connections. That splintering, as well as the marketing of competing journals on the basis of their political positions, in some senses allowed for the very production of a medical politics defined by pedagogical reform.

RADICAL REFORM—THE *LANCET*

During the first three decades of the nineteenth century, at roughly the same time that weekly journals were making themselves the monitors of medical teaching, they also became the conveyors of proposals regarding reform in medical education and a conduit for calls for improvement. These proposals became numerous and their proponents very vocal by the 1820s, aided by the dialogue and audience generated by medical periodicals.[37]

The *Lancet*'s founder, Thomas Wakley, proposed that British medical education and research be refashioned after the French model, in which medical sciences like experimental physiology, pathology, and morphology were being developed in centralized institutions with a salaried faculty. This movement had been born in the aftermath of the Napoleonic Wars. As tensions between Britain and France relaxed, more students sought to complement their medical education at home with study in Paris.[38] Some returned convinced of British medical superiority, but others were impressed by the experimental physiology of François Magendie, the surgical accomplishments of Baron Guillaume Dupuytren, and the new stethoscope and auscultation methods of René Laennec, as well as the superior financial circumstances of those teaching medicine in Paris.[39] For example, in an 1832 editorial review in the radical *London Medical and Surgical Journal*, written by an admiring Londoner, French medical science stood as an ideal to strive for and a bit of a reproach to the British medical scientists and the populace they served:

> The steady and rapid progress of the medical sciences during the present century, has been in a great measure owing to the minute and systematized researches of the French school of pathologists. Much of their superiority over ourselves may be fairly ascribed to the wisdom of their hospital regulations, which enforce the universal inspection of the dead, and to the entire emancipation from prejudice on this subject of the popular mind.[40]

This sort of critique, suggesting that the British were not doing enough to advance medical science, was common in radical circles.[41] The *Lancet* regularly included reports from the French hospitals Hôtel-Dieu and Hôpital de la Pitié in its reports of hospital cases, which was not the case in the moderate medical journal the *London Medical Gazette*.

For those who advocated emulating the French, or just overhauling the British system, the *Lancet* was the primary vehicle for advancing their position. In the *Lancet*, Wakley and others wrote scathing critiques and mocking reviews of members of the establishment and more conservative surgeons and medical practitioners, referring to those who ran the Society of Apothecaries as "whacks" and the Royal College of Surgeons as "Voracious Bats."[42] One letter-writer, complaining of nepotism in the Royal College of Surgeons, wrote to the editor of the *Lancet* in 1829, "Constituted as the College is, Sir, how can we expect better things? Take, for instance the manner in which vacancies are filled in the council; and let me ask, whether that member who has most influence with his colleagues will not take care to introduce his immediate friend or relation, without any regard to his talent or qualifications as a professional legislator."[43] Such complaints were among the *Lancet's* standard fare — extensive reform of a system was touted, alongside criticism of a group of established London leaders as hopelessly corrupt and profit driven.[44] In addition, the *Lancet* advocated reforms to the anatomy laws and for the advancement of the general practitioner. In doing so, it argued for the destruction of professional divisions: "The profession of medicine in the metropolis," another letter proclaimed, "is far too much subdivided and portioned out into different departments; and it is to this cause that we must attribute the paucity of those who possess that comprehensive knowledge of its various and complex branches, which, though difficult to attain, is, in itself, the best reward of its cultivators, and is the only means of raising them to that rank in society which ought to be the ambition, alike of the surgeon and physician."[45]

Many of the journal's contributors were general practitioners, those who practiced midwifery, surgery, medicine, and pharmacy, and who could therefore find no license to legally support their general practice (a practice upon which most of the population relied for health care). Apothecaries could not be licensed if they dispensed both advice and drugs, surgeons were not allowed to practice midwifery, and licensed medical doctors and surgeons were not allowed to practice each other's trade.[46] In practice, a number of these general practitioners did hold licenses as apothecaries or surgeons, but their all-encompassing work remained largely unrecognized officially.

The *Lancet*'s contributors also discussed the cost of a medical education in London, focusing particularly on the practical, hospital-based education of which conservative reformers were most proud. In a fairly typical passage, one editor carped:

> Chemistry is taught him [the student] demonstratively; botany is learned by him in the midst of those objects with which it is intended to make him acquainted; anatomy is subjected to his senses in every possible shape and form; each, in short, of the other sciences is taught him in daily courses of three, four, five, and six months, and often at an expense which he can afford; but that science to which all the others are but subsidiary, is not taught him at all. While an elaborate course of instruction in any of the sciences is given for *four guineas*, the inspection of the patients in a hospital, without a word of instruction, costs him nearly *thirty*![47]

Hospital instruction was one of the best sources of income for teachers of medical subjects, and those teachers thus charged students high prices for access to the hospital and that crucial experience of seeing a large sampling of diseases. Students who paid those prices often felt that their access was inadequate. Some complained that they were not allowed to read through the notes on each patient (rules at charity hospitals varied and some prohibited students from seeing the notes on patients as an attempt to preserve patient privacy), while others felt that clinical lectures were not truly clinical in that they did not make enough use of the patients themselves.[48]

Wakley and his followers, some of whom were private school teachers in London or prominent advocates for general practitioners, held meetings at the Crown and Anchor in 1831 to establish something that Wakley called the London College of Medicine.[49] It was one of their more ambitious attempts to reform London medicine, involving radical social change. Their idea was to do away with the existing licensing bodies, the Royal Colleges of Physicians and Surgeons and the Society of Apothecaries, and to establish in their place a democratic institution unifying the three professions of medicine. This London College of Medicine would remedy much of what Wakley saw as wrong with British medicine. Its officers would be elected annually. It would certify practitioners cheaply. With a vocal leader, a meeting place, and a publishing outlet to unify them, radical reforms presented the picture of a well-defined political group with a coherent platform.

CONSERVATIVE REFORM: DEFINING BRITISH MEDICAL TRADITION, OPPOSING THE FRENCH

Arguments for conservative reform (reform that would *conserve* a British tradition) were more dispersed and varied than those of the radical reformers. They included calls for revision of elements of British medicine—often licensing requirements or structures of the College of Surgeons or College of Physicians—while keeping the essence of London (British) medicine and its social structures intact. Conservative reform embraced a "British" style of medicine and science, while simultaneously contrasting it with "French" or "continental" medicine. In doing so, conservative reform created and defined a British medicine that its proponents described as traditional. Calls for conservative reform emphasized moderation, holding out the specter of the French revolution as a reminder of the perils of any kind of social reform gone too far and of embracing anything French. One such passage embodied sentiments present in many similar articles in the *London Medical Gazette*: "Reform is the wish of many—revolution is the desire of the demagogue alone."[50] Still, conservative reformers hardly formed a cohesive social group and certainly were not uniformly aligned with broader political parties and ideological politics.

Conservative reformers' opposition to using continental medical sciences as the basis of new institutional curricula or requirements for practice was not simply because its advocates did not practice such sciences themselves, but because (they argued) it was necessary for the British to establish their own system, one that was particularly suited to their countrymen.[51] Though the groups were distinct, the rhetoric surrounding such calls for a particularly British kind of reform echoed that of conservative political reformers, who agitated for parliamentary reform between 1780 and 1830.[52] These political reformers argued for British exceptionalism in politics and thought that reform could prevent revolutionary upheaval of the kind found on the continent.[53] Many sought a middle path between reactionary and revolutionary action. The argument that appropriate reform could serve to prevent revolution was made by Edmund Burke as early as 1790 and was very similar to that embraced by conservative medical reformers with respect to the medical world.[54]

Medical reform rhetoric tapped into the notions of both British exceptionalism and reform as an antidote to revolution that were a part of this broader rhetoric of reform. An 1828 editorial in the *London Medical Gazette* captures such resonances, saying, "We have in a former paper expressed something like an opinion, that if medical education in England is not absolutely the best in the world, it is perhaps the best for us: in saying so, we are fully sensible that many improvements might be suggested, but then those improvements and alterations must all be made in the spirit of the English system, (if we may so term it) and according to the feelings and principles still cherished in this country."[55] Another editorial, written in 1832, suggested that proponents of the French system "forget the very latitude in which they live—they overlook the existence of a whole system intrinsically dissimilar—they allow nothing for national peculiarities—or, with a complete ignorance of human nature, they would attempt to drown them."[56] But while we may see reflections of Burkean arguments in conservative reform rhetoric, medical politics was also local, based on professional affiliations, and the arguments took on the particularities of the social world of medicine. Anti-French rhetoric was employed in the service of specifically medical arguments, arguments holding that British, and particularly London, medicine was worth preserving and improving. Such rhetoric helped to construct a British nationalism and ideas about British tradition through medical debates.[57]

So what made a system "British" and what made this British system superior in the eyes of these opponents of French Science? Londoners claimed that, institutionally, it was competition. They prided themselves on the many private schools run by the best practitioners. By 1800, almost half of provincial practitioners had been educated in London (and, of course, an even larger percentage of London's practitioners).[58] With approximately fifty courses and twenty schools or hospitals, students were supposed to be able to find the best teachers, and those who provided the newest theories and practices. In an 1830 editorial in the *London Medical Gazette*, one Londoner wrote, "While we here enjoy all the facilities of instruction and mutual co-operation that zeal and competition can supply—with all that freedom and independence so characteristic of the nation—the French faculty are entirely under the controul of government; in fact, under the surveillance of the police."[59] Similarly, Charles Bell stated in his introductory lecture at the opening of the first university in London in 1828[60] that he hoped that the establishment of the London University did nothing to discourage teachers at the private schools.[61]

That emphasis on competition extended beyond London to other cities in Britain. Even cities with established university medical schools like Edinburgh had a variety of private schools offering courses in medicine. Students would create their own curriculum by attending a mixture of university courses and courses at any number of private institutions.[62] When the faculty members at the University of Edinburgh were not good teachers, as was the case with Alexander Monro Tertius,[63] the private schools of Surgeon's Square did a better business than the University itself.

As in Edinburgh, medical and surgical lecturers in London derived their income from individual students and therefore needed to be popular in order to make a living. Londoners extolled the virtues of this competitive environment, declaring it a way of ensuring a superior education in which the best teachers taught subjects that students cared about most.

Teachers often saved their discoveries for the classroom in order to attract more students by offering access to the latest developments in medical science. Such was the case with Bell's discovery of separate roots of motor and sensory nerves, which was taught long before it was published.[64] In those cases, the classroom itself became the seat of British medical "science." In an 1809 letter to his brother, Bell drew a distinction between medical men who earned their income through practice and those who earned it in the classroom, saying, "My means of being known are through my books and pupils: I retain my consequence by preferring science to practice [. . .]. Those with whom I stand contrasted are making perhaps 9000 pounds a year. What does that imply? The pursuit of science, perhaps? Pooh! Pooh!"[65]

Conservative reform in British medical education also entailed the "tradition" of joint education in surgery and medicine, something that might be regarded as French by the historian but that was carefully couched in

British nationalism by Bell's contemporaries. Bell spoke in favor of conservative reform when he counseled his students in 1835 to know medicine and surgery and to base their practice on a knowledge of anatomy, saying, "Your long list of certificates you must have; but I conjure you to act as if anatomy, and such uses of anatomy as you see in hospital practice, were the business of your life in London, and not to be satisfied with learning to answer such questions as may be put to you at any board."[66] Some of this generalism can be attributed to the fact that most medical practitioners, whether certified in London's classrooms as surgeons or apothecaries, would, to a great extent, serve as general practitioners, dabbling in both surgery and medicine as their patients required.

Comparisons drawn in British medical journals between British and foreign medicine reveal additional ideals of the British traditionalists. When comparing themselves to the French, British surgeons and physicians tended to emphasize their abilities as practitioners and the corresponding strength of the practical education their students received in hospitals. In 1831, the *London Medical Gazette* proclaimed "the efficiency and superiority of the practical character of our own [approach]":

> Upon the rational empiricism of medicine as founded on the study of disease in the great book of nature, has the practice of it in this country been able to bear triumphantly the comparison with its condition in France; there, science studied in the abstract, and theories ingeniously devised, give evidence of the speculative character of that school, whilst the smaller proportion of deaths in England prove as incontestably the efficiency and superiority of the practical character of our own.[67]

British doctors and surgeons emphasized that they spent much of their education walking the wards of hospitals or sometimes in apprenticeship, and the stress in such systems was on therapeutics. The contrast was drawn negatively. According to Adolph Muehry, a British physician and surgeon, "The French physician [. . .] thinks more of the disease than the patient."[68] And one student studying in France wrote home describing the French by saying, "Indeed, they seem to think that the perfection of medicine consists not so much in keeping patients alive as in foretelling with precision the appearances which will be found after death."[69] That reputation was coupled with the association of French medicine with experiments that were often depicted as either establishing facts already known through experience,[70] or cruel and irrelevant to any kind of medical practice. One medical student, James Macauley, wrote: "In 1837 I attended, along with Edward Forbes and others known to you, the class of Magendie; at least we went to some of his lectures. The whole scene was revolting; not the cruelty only, but the 'tiger-monkey' spirit visible in the demoralized students. We left in disgust, and felt thankful such scenes would not be tolerated in England by public opinion."[71] The experiments conducted were depicted as both cruel and revolting *and* as demoralizing to students.

By contrast, one 1833 review of Bell's newly published Bridgewater Treatise, *The Hand, Its Mechanism and Vital Endowments as Evincing Design,* said of Bell: "although in some works we see him designated by the supposed proud title of 'experimentalist,' his experiments have been few, and only confirmatory of deductions previously drawn from studying the course of nerves . . . ," adding with approval that his limited experimentation was "as strictly under Bacon's definition of induction, or 'experience,' as the most multiplied course of experiments . . . although the adherents of the opposite opinion urge that . . . we should 'torture nature,' to make her disclose her secrets."[72]

The rhetorical stress on the practical character of British medicine even extended to the anatomy schools, where the instructors intended their anatomy courses to be "practical." Though dissections were conducted here, as elsewhere, by an assistant, or demonstrator, one advertisement for Bell's Great Windmill Street School of Anatomy proclaimed that "Mr. Bell will continue to visit students during their operations to point out the application of Anatomy to practice."[73] Bell's advertisement claimed that his dissection course was practical in the sense that the anatomy taught was always related by the instructor to clinical practice.

Conservative medical reform involved defining British medicine in opposition to French, or continental, medicine and opposing radicals who advocated French medicine in Britain. The fundamental features of British medicine, according to this version of British tradition, were various: the competitive and open nature of British medical courses; the integration of some medical and surgical training in order to prepare students for general practice; the practical and therapeutic character of a British education; and the absence of unnecessary or unnecessarily cruel vivisection experiments. Benjamin Brodie's 1838 *Introductory Discourse on the Studies Required for the Medical Profession,* delivered to the students of St. George's Hospital at the commencement of the academic season, contains a passage that highlights nicely the ways in which this rhetoric defining British medicine as different from, and superior to, continental medicine functioned. Brodie wrote, "It is one advantage arising from the peculiar constitution of the London medical schools, that, with few exceptions, the instructions which you here receive, have, in a greater or less degree, a tendency to practice. The ambition of the teacher of Anatomy is not limited to success in his present vocation. He looks forward to the time when his profession as a Physician or Surgeon will elevate him to fame and fortune."[74] In this passage, Brodie combined those things that conservative reformers thought made a London education great: its emphasis on practice and the competitiveness of its teachers. Brodie continued:

> I have no doubt that the praises which are bestowed on some of the continental anatomists are well founded: that there are universities in which the Anatomical professors, devoting their whole time, and industry, and intellect, to this one pursuit, explain the mysteries of minute anatomy at greater length, and with more precision, than the teachers here: but, nevertheless, I assert, that ours is the better method with a view to the education of those who wish to become, not mere philosophers, but skilful and useful practitioners.[75]

Brodie rounded out his discussion of the strengths of a London education with a favorable contrast to the French. This sort of nationalistic rhetoric and juxtaposition of the British and French styles of medical education was a resource for conservative reformers in disputes with their more radical countrymen over the substance of reform.

While conservative reformers extolled the virtues of a British medical education, they also recognized a need to improve elements of that education. Their proposals for reform often came in the form of institutional change. Sir Charles Bell testified twice about unjust promotions within hospitals before the 1834 Select Committee of the House of Commons, which was appointed at the instigation of Thomas Wakley "to inquire into and consider of the laws, Regulations and Usages regarding the EDUCATION and PRACTICE of the various Branches of the MEDICAL PROFESSION, in the United Kingdom."[76] He was called to give this testimony because of a pamphlet that he had written in November of 1824, entitled, "A Letter to the Governors of the Middlesex Hospital, from the Junior Surgeon." This pamphlet presented a strong critique of the hierarchy in British medicine, but also Bell's deeper concern for the structure of British medical education.[77] Bell maintained that "the situation of Physician or Surgeon to an hospital, should be a reward for professional merit." Nonetheless, in practice, young physicians who received such a hospital post stayed only as long as it took to learn their profession in preparation for private practice. Then, "whenever their private patients promise them a livelihood, they . . . leave the hospital to the next candidate for the notice of the town." Bell declared:

> In this scheme of forming physicians, there is no provision for the improvement of science or for the records of practice; neither stimulation nor reward is held out [. . .]. In the course of a few months a young gentleman is a Student, a Member of the College, Physician to an Hospital, and Teacher of Medicine. It would be well if he were to proceed at this rate; but a few private patients withdraw him from his public duties, and he is influenced by that notion which prevails so extensively in London, that to be otherwise employed than moving about in a chariot, is to declare his incapacity.[78]

He was objecting specifically to the promotion of a new fellow of the College of Physicians over a more experienced physician at the Middlesex Hospital, but he was also objecting to the idea that a lucrative private practice was considered the physician's end goal. Teaching and medical science suffered as a result of such goals, because men of experience never stayed on to promote the development of new knowledge.

These sorts of critiques—of unjust promotions, impenetrable hierarchies, and unworthy elites—constituted a rhetoric widely shared and present across the British educational system and political spectrum. In 1800, the physician and reformer Thomas Beddoes had complained, as quoted by Roy Porter, "of its 'system of hereditary professorships' which showed 'every reputed disadvantage of hereditary monarchy, and not one of its advantages,'"[79] while in 1840 an editorial from the *Medical Times* of London was still talking about weeding out "the absurd system of HEREDITARY PROFESSORSHIPS."[80] The question was simply what to do about it. Radicals favored the immediate leveling of professional hierarchies, whereas conservative calls for reform involved the gradual alteration of licensing requirements to reflect the realities of practice and the systematization of a curriculum, acts that would themselves tend to diminish the significance of hierarchy over time.

Conservative proposals for reform were carefully crafted to improve British medicine without restructuring it or undermining its foundations. They emphasized the need to revise the existing Colleges of Surgeons and Physicians and Society of Apothecaries in order to make them more democratic and just, but not to replace the institutions. Reformers proclaimed the need to codify a set of courses necessary for the practice of medicine and to broaden that set so that surgeons, apothecaries, and medical doctors would be equipped for the sort of general practice that they were likely to engage in once licensed. Such proposals claimed to accentuate valued elements of a British medical education: if competition was meant to produce the best education, then making the governing bodies of medicine more democratic would only improve them; if a British medical education was the best in the world because it was practical, then creating a system that reflected the realities of general practice and increasing access to bodies for practicing dissection would only make it more practical, amplifying its superiority. In 1831, these competing reform ambitions were showcased in a dispute over the formation of Wakley's London College of Medicine.

LONDON COLLEGE OF MEDICINE: THE RHETORIC OF REVOLUTION AND REFORM

On May 7, 1831, Thomas Wakley and a number of his radical reformers held the first public meeting at the Crown and Anchor in London to set out the framework for what they called the London College of Medicine.[81] The London College of Medicine was meant to serve as a rival to the Royal College of Physicians, the Royal College of Surgeons, and the Society of Apothecaries, as well as to serve as a licensing and professional body for all medical practitioners: physicians, surgeons, apothecaries, and general practitioners. It was a revolutionary proposal, and if the two platforms laid out in this chapter—those of radical and conservative reform—held sway among particular social constituencies, as the rhetoric of the new weeklies suggested, the proposal should have divided journals, men, and professional groups directly along political lines. It did not.

The proposed structure of the London College of Medicine addressed many of the concerns that were a part of the radical reformers' platform—in particular, the concern that the separate licensing of physicians, surgeons, and apothecaries did not reflect the realities of general practice, and the critique that licenses were granted on the basis of certificates showing attendance at courses and payment to London lecturers and not on the basis of skills performed as a part of the examination. The policies set out in the first meeting of the College decreed that any man who was legally qualified to practice any branch of medicine in England, Ireland, or Scotland, or could produce a diploma from any British university or medical college at the time of the College's founding would be deemed eligible for a diploma (license to practice) without examination. All those who received the diploma of the London College would be considered Fellows within the institution and "Doctors" outside of it. The college would be governed by a senate elected by the general membership. The senate would examine future candidates for diplomas, those examinations would be conducted in public, and candidates would not to be required to produce any certificates of coursework. The exam for the diploma would be conducted over two days: the first day would address the "facts of anatomy and materia medica," and the second day would address the "theoretical principles of physiology, pathology, semiology, surgery, and the practical application of these principles to medicine, surgery, and midwifery." The cost of the diploma would be between three and five guineas and Fellows would be at liberty to practice all branches of medicine or to specialize. There would also be an eleemosynary fund established for widows.[82] This proposal contained a number of drastic changes to the system of medical education, licensing, and governance in London.

The London College of Medicine would have united doctors, surgeons, and apothecaries in one body, leveling the entrenched hierarchy of the London medical scene. It also would have recognized training in provincial schools and hospitals as equal to that occurring in London, as had not been done by the Royal Colleges. It would have been led by a popularly elected senate, would have reduced the cost of a diploma, and would have disrupted the power of hospital surgeons by not requiring certificates of attendance at a hospital. Wakley emphasized several times that its policies had been founded on principles taken from the best examples of medical societies in Britain and *around the world* (the international scope also reflecting the position of radical reformers, who often relied on French examples as they reimagined London medicine).[83] As Wakley himself put it: "[h]ere the power rests with the whole body of the Fellows. No man is to receive the diploma without enjoying full equality in collegiate rights and privileges. . . . The funds of the College will be secure against plunder. Merit will be protected and rewarded. Students will be examined in public, and be thus shielded from the petulance of ignorant bigots . . ." Wakley also alleged "malignant, secret intrigues of rival teachers." His plan was to abolish "distinctions which have so long disgraced the profession," so that "the public will have the infinite satisfaction of knowing that every possessor of the diploma of this College has proved that he is well qualified to practise in every branch of medical science."[84]

Responses to this proposal, which was crafted around the radical reformers' agenda, were, in substance, surprisingly similar across the two different journals, with their supposedly different communities. Readers wrote to the *Lancet* and applauded the new London College of Medicine, pledging their commitment to it, but those same letter-writers who praised the institution also expressed reservations. A number of readers of the *Lancet* wrote letters voicing concern about the extreme nature of the leveling of medical professions that the London College of Medicine would bring, about the proposal to admit initially all who had legally practiced any branch of medicine prior to the College's founding, and about the wisdom of removing requirements for hospital attendance, at the same time that they avowed support for the institution and for Wakley.[85] One anonymous letter-writer, sounding very moderate indeed, wrote,

> Living, as we do, in the age of medical reform, it would be well for us to recollect, that it is less difficult to be convinced of the necessity of reform, than to define the nature of that change which would beneficially and permanently influence the whole medical community. Salutary reform does not, in my opinion, consist in removing every impediment for the purpose of establishing perfect equality; nor is the leveling system carried to a fearful extent, at all compatible with the best interests of the profession.[86]

Another added that "there surely ought to be some distinction made between a strictly professional man, and one who is half doctor, half tradesman; *you* may pass decrees declaring their equality; but will society, for whom artificial distinctions are made, recognize and act upon your laws?"[87] Both letters suggest that readers of the *Lancet* were concerned about keeping out quacks and about overzealous demolition of professional divisions, but they were also particularly enthusiastic about the proposal to merge surgery and medicine (if not necessarily including the drug-selling apothecaries) to create a single diploma, and about the increased openness and inclusiveness, the democratic nature, of the new institution.

Conservative reformers writing both in and to the *London Medical Gazette*, not surprisingly, opposed the founding of the London College of Medicine for its revolutionary institutional solution that threatened to destroy and completely rebuild London medicine, but they also framed their opposition in ways very similar to those of the *Lancet*'s correspondents. One concerned medical man wrote to the *Gazette*, echoing the fears of his counterparts, "If those are to be admitted who were in practice previous to the year 1815 . . . the College would have hosts of middle-aged chemists and druggists, oilmen and grocers, presenting themselves as candidates for the diploma, stating that they were in practice previous to the passing of the Apothecaries' act: and who could deny it?"[88] The writer was alluding to the relatively unregulated state of apothecaries before the Act of 1815 that imposed strict licensing requirements on those who wanted to dispense medical advice, and his concerns were expressed in a typical fashion. Another author wrote, "it is stated that candidates will not be required to produce any certificates whatsoever—an examination being considered quite sufficient: even a certificate of hospital practice is not required—I am sure it is not necessary to occupy the pages of your journal with arguments proving the complete absurdity of this."[89] These statements are remarkable for their similarity to those made by subscribers and readers writing to the *Lancet* who expressed concern about keeping the caliber of professional medicine high.

But in this period, reform rhetoric circulated freely, and those who criticized the London College of Medicine on some grounds also recognized the need to change membership requirements of the Royal Colleges and to make them more inclusionary. Writers to the *London Medical Gazette* suggested, "No man, who is not very blind or very uncandid, will deny that the present constitution of the College of Surgeons is essentially unpopular . . . at variance with the spirit of liberalism which marks the times."[90] Liberalism, freedom, education, improvement, and, of course, reform were watchwords across the spectrum of medical politics. Another surgeon asked of the Council of the Royal College of Surgeons, "Why should the Council remain insensible to the advantages which would arise from a freer intercourse, in their official capacity, between them and the Commonalty, which, without trenching upon their privileges, would increase the importance, the utility, and the character of the College?"[91] And one author captured particularly well the sentiments of many who wrote to the *London Medical Gazette*: while noting that most "reflecting men" believed that reform in British medicine was needed, he cautioned that, nonetheless, "the proposed '*College of Medicine*' is not likely to effect any amelioration or reform in medical polity; first, because many of its proposed measures are objectionable, or even Utopian; and, secondly, because the 'College of Medicine' will not be joined by the most respectable and influential members of the profession, for reasons which need not here be stated." Instead, what was wanted was an "Association for the Improvement of Medical Education and Polity, and for promoting harmony in the profession."[92] The consensus seems to have been that, although Wakley's new College went too far, although it jeopardized medical education and risked letting quacks practice legally, reform was necessary, whether or not it was reform along the lines of the principles that Wakley and his cohort espoused.

But while the echoing rhetoric points to considerable agreement by two competing journals on a number of points, the debate surrounding the London College of Medicine was also revealing of divisions. These are suggested in the passage above, in which the author wrote: "the 'College of Medicine' will not be joined by the most respectable and influential members of the profession." The dispute over the College of Medicine was grounded in a nascent professional politics that divided as much as in a political rhetoric that blurred some of those divisions. What this suggests is that, while the new journals served to create new kinds of communities, those communities had not wholly supplanted social networks that had taken root in a dogmatic and ideological sort of way. The *Lancet*'s audience and leader had formed a social group, unified through in-person meetings, but even their cohesiveness was tested by the more extreme politics of the London Medical College, while the *London Medical Gazette*'s audience formed a much looser network, composed of those with flexible politics who were generally moderate in their reformist ambitions. The journals' audiences disagreed about aspects of the London Medical College in ways that corresponded to professional politics, but their supposed disagreement was articulated through a common rhetoric of reform, suggesting much that was shared by these emerging print communities. Charles Bell, whose claims in a priority dispute were grounded in nationalist, anti-French sentiment and who was supported by an old network of patronage, was working in a changing, if not yet fully changed, world.

Issues surrounding the London College of Medicine were, in large part, framed as having to do with the politics of revolution versus reform—should institutions be destroyed and rebuilt or slowly altered, working with existing structures? One reader of the *London Medical Gazette* set it up that way directly, saying, "I am a decided friend to reform, but not revolution,"[93] explaining: "In reference to this institution [the Royal College of Surgeons] I would say that it admits of, and indeed requires, very considerable reform; but because it requires reformation, is no reason that it should be annihilated."[94] The distinction between revolution and reform carried political implications: it was a risky position to be a revolutionary in early nineteenth-century Britain, where the specter of the French Revolution loomed large and its violence and social upheaval were to be avoided at all cost. The editor of the *London Medical Gazette* made such a point explicitly, saying of the Royal College of Surgeons:

> There are many who may be called *reformers*—that is, who would prefer having a voice in the election of the Council, and who would be glad to establish a right of property in the building, such as should enable them to assemble there to

> discuss professional matters. Of such persons a great majority of the members consist. But of *radicals*—that is, of those who would support the riots got up by Wakley as a fillip to the sale of his papers—there is not one in fifty, and those few are persons of no note, influence, or name.[95]

Professionally, the politics of reform was safe; there were many who advocated reform. When Wakley called the London College of Medicine "a complete renovation of the medical profession in England—an entire remodeling of the statutes relating to medicine,"[96] and said that the medical profession would regret having toiled for many years "under the iron rods of incompetent rulers" when they could have "thrown off the yoke of their debasement,"[97] there were few who wanted to follow his lead in thus courting revolution. Those who sought to do so nonetheless downplayed revolution in deference to reform.

By the end of 1831, the London College of Medicine had simply faded away.[98] This happened in part because even its supporters had reservations about its extreme positions, and in part because Wakley, its main promoter, embraced the rhetoric and acts of revolution in a time when reform was its safer, and indeed more British, alternative, no matter what one's political circle.

CONCLUSION

Between 1810 and 1830, a variety of reformers became increasingly critical of British medical education. Newly emergent medical journals showcased pedagogy, as proposals for the reform of education took shape in their pages. The rhetoric of reform was very much a part of institutional changes that took place in London's teaching hospitals as well as in the erection of a new university in London in the late 1820s, as we will see in the next chapter. In those institutions, reformers debated again whether British medical education should be fundamentally practical and whether it should be cosmopolitan.

Prominent practitioners and contemporaries of Charles Bell like Sir Astley Cooper display the lack of coherence in the politics of conservative reformers. The *Lancet,* calling attention to nepotism in the London hospitals, published an article stating that Cooper had five relatives working in key positions in London's hospitals and that their combined income from students was three thousand pounds.[99] But Cooper cannot be neatly categorized. For example, he maintained friendships with known London radicals

Henry Cline and John Thelwall[100] and went to Paris to see the revolution firsthand. Later, however, he wrote, in what was a fairly typical antirevolutionary call for reform: "A revolution may sometimes be a good thing for posterity, but never for the existing generation for the change is always too sudden and violent."[101] Sir Benjamin Brodie[102] and Cooper worked to obtain a royal sanction for the London Medical and Chirurgical Society in 1834, a recognition that had initially been opposed by the Royal College of Physicians. The Medical and Chirurgical Society brought together individuals from all branches of medicine in the service of the improvement of that science, though it had no ambitions to grant licenses or to otherwise alter the power structure of London's medical world and its founders sought authority granted by already recognized, traditional institutions.[103] In other words, it was a moderate attempt at achieving some of the same goals as the London Medical College.

Medical weeklies' printed responses to various reforms of London's medical institutions demonstrate that professional and political communities were only just being formed, and actors themselves moved fluidly between groups and positions, as was necessary in the small and patronage-driven world of London medicine. At first glance, the medical weeklies in which pedagogical reform was debated seem to show new virtual communities, formed along political lines, built on ideology rather than on local politics. Journals staked out overt political positions in attempts to compete with each other for audiences, even while presenting much the same material, reflecting the same reports of the contents of London's classrooms. But letters from readers suggest that those journals, with their rhetoric of radical or conservative reform, mapped imperfectly on the audiences that they served, and that the incompletely aligned politics of their readership did not yet constitute new social groups and allegiances. Such communities were more likely to take definite shape and a politics of their own through the building of institutions that offered their participants careers that promised to supplant the cultivation of patronage among existing social powers—educational institutions whose development is the subject of the next chapter.

CHAPTER FOUR

London's New Classrooms: London University and the Middlesex Hospital School

> I don't think I have written to you since I began the lectures, establishing a school in the old Middlesex. At least, my spirit and devotion to the art and to the institution to which I am attached, will not be denied. I have delivered six lectures, such as only long experience and study could have produced.
>
> CHARLES BELL TO GEORGE BELL, October 7, 1835[1]

In 1826, after fourteen years there, Charles Bell sold his stake in the Great Windmill Street School (handing it over to his own students Herbert Mayo and Caesar Hawkins), and began to focus on his surgical lectures in the wards of the Middlesex Hospital, where he had been appointed as a surgeon in 1812. Despite classroom successes, Bell remained, as ever, too poor for his tastes, so when he sold his interest in the Great Windmill Street School, he also sold his museum collection of specimens and preservations to the University of Edinburgh to generate extra income, and then began collecting anew (enduring scathing critiques from Thomas Wakley for the lifelong practices of specimen acquisition that supported Bell's enduring object-based teaching practices[2]); Robert Knox, the popular lecturer and anatomist, came down to pack them up.[3] In addition to his usual worries about money, Bell was plagued by other anxieties: as he operated more, both within the hospital and without, Bell began to suffer from a great deal of apprehension about performing surgery. The death of one of his patients haunted him: "I have had a most miserable time since I wrote to you, from the failure of an operation, and the death of a most worthy man. I shall regret it as long as I live. It is very hard, more trying than anything that any other profession can bring a man to."[4] And in February 1826, he wrote again to George almost three years after that, "I suffer indescribable anxiety, so that I vote my profession decidedly a bad one."[5]

Thus, with money but no school, attempting to avoid a career simply as a practicing surgeon, Bell became restless. An unbusy life was not his and he was ready for a new venture. Teaching was always his primary occupation, and having given up the Great Windmill Street School (figure 12), Bell had time again to think about how it might best be pursued. His timing was excellent and his political connections helpful, for in 1827 he joined an old friend and patron, Henry Brougham, MP, at the Freemasons' Tavern for a dinner laying the foundation of a new university in London, one that was meant to reform British education and capitalize on the forms of practical

FIGURE 12. Watercolor of William Hunter's Great Windmill Street School. It was during his time as proprietor of the Great Windmill Street that Charles Bell really established himself and began to act on his ideas about pedagogical reform. Photograph of watercolor courtesy of the Wellcome Library, London.

training in a variety of middle-class professions that London offered.[6] Institutional politics were not Bell's interest, but they were a means to an end for him as a pedagogical reformer.

In the Great Age of Reform, nearly everyone was a reformer. In the medical world, most of those reformers agreed that medicine should be both "scientific" and "practical." London University was founded in 1828 with such ideals in mind. And yet, despite apparent agreement built around the vague principles of furthering practical science, London University seemed to be an institution full of chaos and conflict within just a few years of its founding. The terms meant different things to different people and loosely built alliances fell apart in the context of making practical curricular decisions and running the university. The debates surrounding its birth and the changes it underwent in its early years regarding its medical curriculum help to capture the transformation from a model of medical science that was intended to be particularly British to one that emulated continental rivals and was considered to be universal.

London University was clearly meant to alter the landscape of British education, and medical reform was central to it. Before the opening of the university in 1828, London had no university medical school. Previously, medical education in London was conducted in disparate and varied institutions. It was ad hoc, meant to suit an individual's needs. There were two primary components of London medical education: the private schools of anatomy, which also offered medical courses other than anatomy depending on the lecturer's skills,[7] and the hospitals, which offered the opportunity for students to "walk the wards" and to attend clinical lectures.[8] London University was meant to combine the best from each source. All agreed that London University should offer an alternative to the classical learning of Cambridge and Oxford, but reformers debated just *how* medical education should be refashioned at the new institution.[9]

Charles Bell was among the twenty-four founding members of the faculty of the University and gave the first address at its medical school. In letters leading up to the founding, he wrote to his brother of being hesitant to join, saying that the University would need a charter and the right to grant degrees if it were to be worthwhile.[10] But he did join the faculty in the end, and in his first address spoke of the potential value of the medical school. "If I value highly the influence of this great establishment," he said, "it is because I have been long engaged in teaching, and have experienced all the difficulties of forming a medical school. For obvious reasons, London must continue to be the principal school of medicine; but whilst there are many favourable circumstances, there are also many unfavourable to regular study; and it is now to be demonstrated that it is possible to retain that which is favourable, and to avoid the defects."[11] Bell went on to situate London University clearly amidst London's professions and trades, describing the training that the university would provide as requiring "regular study" but also being "more practical [than other universities]," and recognizing all the attendant benefits and difficulties of the medical experience that London could provide.[12]

London University, founded in 1826 and opened for classes in 1828, was intended to serve merchants and members of the professions of medicine and law and to conduct courses in English, without the emphasis on Latin and Greek found at Oxford and Cambridge. In doing so, it would use the "local advantages in the metropolis, for connecting the theoretical with the practical parts of these branches of knowledge, which cannot be equally enjoyed in any provincial situation."[13] Council members hoped that the university might help to rectify a system of the professions in which the majority of practitioners of both law and medicine had not graduated from universities, and thus provide those practitioners a more thorough and consistent education.[14] The University was meant to be flexible and utilitarian. Its founders, in an attempt to make it adaptable to the changing sciences and the needs of students, instilled in it a "perfect freedom of competition,"[15] providing faculty with only small salaries, the bulk of their income coming from student fees, much like the system at the University of Edinburgh. As it was explained by Thomas Macaulay in the *Edinburgh Review*, "Under such a system [. . .] whatever language, whatever art, whatever science, it might at any time be useful to know, *that* men would surely learn, and would as surely find instructors to teach. The professor who should persist in devoting his attention to branches of knowledge which had become useless would soon be deserted by his pupils."[16] Good teachers who taught useful courses would prosper.

The University's founding was largely the work of the Edinburgh lawyer, Whig, and reformer Lord Henry Brougham, whose interest in education is exemplified by his involvement in the Society for the Diffusion of Useful Knowledge (SDUK), the London Whig association intended to provide scientific and practical knowledge to the middle and working classes.[17] Brougham was also to be a staunch supporter of the Reform Act of 1832 and believed that reform, coupled with education in the sciences, would help to improve the lot of the middle classes and the overall welfare of society. When it opened, London University, widely known as "the godless institution of Gower Street,"[18] had no religious requirements. According to its widely distributed prospectus, the University was meant "to bring the means of a complete scientific and literary education home to the doors of the inhabitants of the metropolis, so that they may be enabled to educate their sons at a very moderate expense and under their own immediate and constant superintendance."[19] This would distinguish London University from its ancient competitors, for, as the prospectus declared, "It is known that a young man cannot be maintained and instructed at Oxford or Cambridge under 200£ or 250£ a year while the expenses of many very far exceed this sum; and the vacations last about five months in the year. The whole expense of education at the London University will not exceed 25£ or 30£ a year."[20] This moderately priced education was meant, above all, to be practical in nature.

Even the allocation of space, set out in the original prospectus, was evidence of London University's commitment to a practical, useful education. The initial plans for the university building called for four lecture halls of varying sizes. The prospectus makes clear that the largest rooms would go to courses in "Anatomy, Surgery, Midwifery[,] Chemistry, Materia Medica, Chemistry applied to the Arts[,] Mechanical Philosophy, Geology and Mineralogy, Mechanical Philosophy applied to the Arts[,] Nature and Treatment of Diseases, Physiology, English Law."[21] The biggest rooms, it seems, would be reserved for subjects with direct and clear practical application. Sciences would be taught in those rooms if they were to be applied to the arts—only the most practical of medical and surgical subjects, along with a practical course on English Law, would be large enough to require such space. Smaller lecture rooms would be used for courses in such subjects as Latin, Greek, mathematics, and jurisprudence, while the smallest rooms would be saved for classes on subjects like history, logic and philosophy, botany, and zoology. These classes, designated for the smaller rooms, reflect the anticipated lower enrollments in courses that largely lacked clear utility in the professions or trades. The plan reflected the anticipated "utility which prevails in the class for whom the Institution is peculiarly destined."[22]

In medicine, "useful" and "practical" knowledge was cultivated at the bedside. The Diploma of Master of Medicine and Surgery[23] required that students acquire certificates of honor in classes on the "practice of medicine, anatomy, physiology, surgery, midwifery and diseases of women and children, materia medica, botany, chemistry, and anatomical demonstrations and dissections as well as attending the medical practice of a hospital containing at least 100 beds for 12 months; and surgical practice in an hospital meeting the same requirements."[24] The extensive hospital practice required attests to the importance accorded it by the founding members of the medical faculty. Bell, who assumed the professorships of surgery and physiology at the outset, was particularly insistent on the importance of hospital training, saying in a letter to his brother from 1828, "There is a plan of uniting London University and the Middlesex Hospital. I have calmly looked to this as the only thing they can do. . . . An hospital is necessary for our curriculum."[25]

By teaching medicine and surgery side by side to all its students, London University's council was catering to a group of general practitioners who were becoming more prevalent in London and the provinces. Apothecaries had the most ill-defined position in the medical world, often functioning as general practitioners and druggists, and were the lowest ranked of these professional groups, governed by the Worshipful Society of Apothecaries.[26] They had the most to gain from the sorts of reform that offered to level professional hierarchies and constituted a large part of Thomas Wakley's radicals. Surgeons, governed by the Royal College of Surgeons, a largely conservative leadership, were of a higher rank and social standing than apothecaries, as a rule, but a distinctly lower one than medical doctors.[27] English medical doctors earned their MDs at Cambridge or Oxford or else abroad (including Scotland). The Royal College of Physicians served as their governing and examining branch. Although men within London were forbidden by the corporations from crossing disciplines (surgeons could not practice medicine, and apothecaries could choose either to dispense drugs or to provide medical advice, but not both), in practice most men outside of London were apothecaries and also operated as "general practitioners."

Apothecaries constituted the first branch of medicine to undergo formal educational reform in the nineteenth century. The Apothecaries' Act of 1815 required that apothecaries possess a license from the Society of Apothecaries in order to practice (exempting men who were already practicing as apothecaries). Formal qualifications for a license required courses in anatomy, botany, chemistry, materia medica, and physic, six months of hospital experience, and apprenticeship.[28] This new licensing helped to certify them to practice as general practitioners and largely capitalized on and made systematic coursework that was already being pursued in London.[29] The course offerings of London University were built around these requirements, as well as the interests of surgeons who planned to work as general practitioners. London University and its early proponents were working to reform the sort of medicine that was practiced in the majority of Britain, providing a systematic education to the general practitioner.

According to those founding reformers of London University, who sought to offer cheaper and more practical classes, the apothecaries and surgeons who engaged in general practice were not well served by English universities and had a reputation for seeking the bare minimum of training before beginning to practice.[30] London University was meant to remedy that. In other words, it was meant to rectify a problem particular to the structure of the British medical profession—that the professional group of surgeons and apothecaries (general practitioners) who treated the majority of British patients had little systematic education and no universities built to suit their needs; it was born as a particularly British, perhaps particularly metropolitan, institution.

It was essential that such an education be a formal and officially sanctioned one in order for it to remedy the historical institutional deficiencies in training for general practitioners. Bell initially kept the new institution at arm's length, writing to his brother in 1828, "I have avoided seeing any of the University people, and mean to keep myself to this: 'Procure us a charter, and a power of giving degrees. Do this and success attends us; fail in this, and the fault is not in us the professions.' My determination is not to enter the walls as a professor until this be done . . . it is the sine qua non, it is necessary to general success—it is necessary to enable me to do good."[31] According to Bell, British, and particularly London, medical educational structures needed to be reformed in order to accord with the realities of practice. To attempt to reform medicine and its hierarchies gradually through new educational institutions and credentialing, as Bell advocated, rather than directly attempting to undo and level professional hierarchies already in place, was to reform without revolution—to reform in conservative rather than radical fashion, as Bell himself favored.

BATTLE OVER REFORM: FROM EDUCATING THE GENERAL PRACTITIONER TO IMPORTING A COSMOPOLITAN MEDICAL SCIENCE: LONDON UNIVERSITY, 1828–35

Between 1828, the year of its founding, and the time Charles Bell left the school in 1830, London University had taken shape, answering London's needs with a cosmopolitan solution born of reformers whose intentions were radical. But in its nascent stages, London University's metropolitan character and the focus on educating a general practitioner made the new university the right sort of place for a man like Charles Bell, who was a great proponent of systematizing medical education by making it comprehensive and by grounding it in anatomy and practical therapeutics.

Bell wrote explicitly about teaching as the best way of improving a science. A systematic education, grounded in anatomy and hospital practice, seemed to underpin and perhaps even to constitute the sciences of medicine and surgery for Bell. In his book *Institutes of Surgery* (1838), published later from Edinburgh, Bell laid out a plan of instruction that encapsulated an idea of a medical curriculum that he had developed over his career, recommending that students have a background in natural philosophy (of the sort that Bell himself bemoaned he'd never had[32]) before beginning their study of medical subjects; that they begin with anatomy, practicing dissection frequently; that they then add some form of mechanical exercise such as drawing or anatomical preparation; that they dissect always with reference to the living body, of which they should acquire a knowledge in the hospital, where they should observe the body and how much the natural constitution can bear; and that they also acquire a knowledge of the medical treatment of surgical diseases (diseases that might involve surgery). Finally, Bell wrote, "clinical instruction is the last and best stage of this laborious course of study: and to maintain his spirits and perseverance during it, the student must look to the noble consequences, the power which knowledge places in his hands."[33] The sciences of medicine and surgery would be improved through such a systematic training.

Bell had hoped that such training would be provided by London University, and he and his fellow reformers intended to make the curriculum comprehensive, affordable, and particularly suited to the general practitioner. There were innovative courses proposed at the University, each with an important practical component: John Connolly proposed a course of instruction in mental disorders and recommended that students be allowed to observe an asylum, and John Hogg opened a dispensary where he conducted postmortem examinations for the benefit of the students.[34]

While those teaching medical and surgical courses were, to some extent, competing for students and student fees, they had hoped that, as Bell put it in his lecture given at the opening of the University, "great advantage and satisfaction [would] result from a combination of learned men, each active in his own sphere, whilst all combine for the greater object [. . .] the improvement of science and literature."[35] But the new university's faculty proved to be less harmonious in its pursuits than Bell and the others on staff had hoped.

Staff appointments were initially determined by a council whose members were themselves businessmen, politicians, and lawyers, and which paid little heed to philosophical consistency in its early appointments.[36] The founding council of twenty-four men included six Members of Parliament and seven Fellows of the Royal Society, but only one physician; their expertise was not in medical pedagogy. From the outset there appeared to be problems with administration and organization of classes; for example, anatomy was taught (under classes of different names) by at least three different professors—Bell, Bennett, and Pattison—and each complained in public about the activities of the others. In the *London Medical Gazette* of 1830, an editorial pointed out that the "three were lecturing in the same classroom on the same subjects, with the same preparations put upon the table, three successive times in the same day."[37] Debates between those who had professional differences took on a personal tone,[38] and arguments became particularly vitriolic when they involved the professors' pay. The pay arrangements for faculty, arrangements that depended on student fees for particular classes, caused controversy.[39] While professors also disagreed philosophically about the content of medical education in ways described below, those disagreements coexisted with disputes that were local, professional, and financial.

The competitive aspect of London medicine seems to have been at the root of a dispute that had significant pedagogical implications: the dispute over a hospital. The council of the University declared in an 1827 statement explaining the strengths of its proposed arrangement that one of the advantages of London was that it presented opportunities to combine theory and practice. To that end, "an hospital capable of containing a sufficient number of patients to afford opportunities of clinical practice, both medical and surgical, and of illustrated lectures" was to be provided, "as an essential requirement of a medical school."[40] Initially, it looked as though that hospital would be the Middlesex Hospital, as both Bell, who occupied the Chair of Physiology and Surgery, and Dr. Thomas Watson, the Professor of Medicine, had appointments there. Bell was in favor of such a union. But members of the University's council were less sure that this was desirable. George Birkbeck, the only physician on the council, for example, wrote of the incorporation of the Middlesex, that "it can benefit nobody but Charles Bell."[41] Surely he meant that such an arrangement would be to Bell's benefit financially. Bell subsequently suggested a link with St. George's Hospital in addition to the Middlesex, allowing students to be admitted to clinical courses at either institution, but the University remained unaffiliated. The lack of a hospital destroyed the hopes of some of the conservative reformers who were touting the University as the place to build a new practical science of medicine—for Bell, training in medicine could be neither practical nor systematic without a hospital. But while radical reformers were by no means opposed to hospital medicine (which was a hallmark of nineteenth-century Paris medicine), they certainly did want to avoid giving support to hospitals entrenched in the London College patronage system.

In May 1830 Bell threatened to resign unless the medical school was remodeled to eliminate the overlap in teaching and to include a hospital. Just a few months later, in September of the same year, he resigned the University's Chairs of Surgery and Clinical Surgery but retained its Chair of Physiology. At the same time it was claimed that he had posted a notice at the Middlesex Hospital saying that he no longer supported the university medical school, at which time the Council asked him to leave.[42] Charles Bell's resignation was highly publicized (all the more so because he was rumored to have resigned, and in fact did submit his resignation several times between that spring and fall, before he actually left the University, mid-term, in 1830), but it was only one of many resignations in the early years of London University.

By the beginning of 1830, the stage was already set for Bell's later departure. In mid-March, conflict had boiled over, and Bell wrote of the fighting at the University, "We had rather a disturbed meeting of the proprietors in the University . . . I have said to the Council they must stop all this, nor allow the professors to write about the University to the public, and through the public newspapers,"[43] hoping to tamp down at least its expression in print.

For Bell, and perhaps his medical colleagues, the promise of London University soon diminished as it became clear that those two crucial ideals of his—those of the scientific and the practical—were not shared, or at least not shared in more than name alone. Bell wrote in his final letter of resignation from London University, "It is impossible that medicine, as a practical science, can be taught without a constant reference to the chambers of the sick, any more than chemistry can be taught without apparatus, botany without plants, or anatomy without bodies."[44] This passage illustrates one aspect of the practical nature of medical science that Bell discusses—medicine was practical in that it was "hands-on"; it was practical as opposed to theoretical. Both conservative and radical reformers would have seen value in practical, hands-on sciences and in moving away from the Latinate theoretical medicine taught at Oxford and Cambridge, but while radical reformers sought to import materialist experimental sciences like physiology from France, for conservative reformers, "practical" meant teaching subjects that were directly related to patient care and that would bear an immediate relationship to the *practice* of medicine and surgery, abstaining from teaching subjects that did not have such a relationship. Medical science, for conservative reformers, required a sort of instrumental effectiveness borne out in general practice.[45] It was a science based in hospital experience. The practical science of surgery, according to Bell, depended on an extended, comprehensive course of such study.

The *London Medical Gazette* in 1834 could paint a picture of educational turmoil at London University, saying that most "respectable men" had resigned "in despair or in disgust"—"the Warden and the Professors of Medicine, or Surgery, of Anatomy, of Medical Jurisprudence, of Clinical Medicine . . . &c. &c. have all changed within an inconceivably short period."[46] The resignations within the medical faculty were partly a result of disputes over turf, of local and professional politics, but they were also tendered by men like Bell who felt that London University was not doing an adequate job of maintaining the virtues of, much less advancing, British (which is to say London) medical education and its strengths in practical training. The rhetoric of conservative reform was built on the notion that British medicine and medical education were superior because they were fundamentally practical.[47]

While radical reformers certainly would have considered elements of their system of education to be practical as well, conservative reformers had a tendency to emphasize hospital training (objectionable, at least in its London instantiation, to radical reformers because the hospitals were controlled by the conservative establishment) and, perhaps more problematically for radical reformers, to deemphasize sciences like experimental physiology that seemed to have no relevance for therapeutics and to be importantly tied to the Continent and French ideas.

Because of his sudden departure, Bell wrote a letter to his students when he left, a letter that was published in the *London Medical Gazette*. In it he wrote: "To those who know how little I value physiology, in the common acceptation of the term, it will be a proof of my desire to see the experiment of the new school fairly tried, that I submitted to be called professor of that science (if a science it be). . . . You are aware that the subjects on which I lectured were the higher departments of anatomy."[48] With the term "physiology" standing in here for continental or experimental physiology, Bell's letter suggests a philosophical difference between him and some of his colleagues that went beyond the question of a hospital: Bell wanted to maintain a British anatomical tradition in the medical sciences. This British tradition, as defined by Bell and other conservative reformers, was always defined in distinction to a "French style" that was seen as callous and materialist.

London University was increasingly becoming a home to radicals like those at the *Lancet*, who, in 1829, could recommend only London University to students. After all, other medical schools, places like St. Bartholomew's Hospital School, "profess[ed] to be a complete school of medicine and surgery, but there [were] no lectures on either *comparative anatomy* or *physiology*."[49] Such subjects were synonymous with a French style. Conservative reformers like Bell accused the radicals of having brought the sort of materialist, politically and scientifically revolutionary, and (most importantly) nontherapeutic sciences to London University that they associated with the French.

Dr. John Elliotson was appointed in 1831 to the professorship of the Principles and Practices of Physic. In his introductory lecture he remarked that, when teaching at St. Thomas's hospital school, he had found that a former hospital-school colleague, with whom he was on the "very best terms," had "in the complete course which, according to our arrangement, he delivered during the season, and I in mine, upon the same subject,—inculcated both principles and practice so diametrically opposite upon almost every subject, that, to use his expressions, 'we must have differed toto coelo, both as physiologists, pathologists, and therapeutists, in many most essential points of doctrine and practice.'"[50]

Elliotson did not elaborate on the points of difference between himself and his hospital-school colleague. But the fact that he spoke with pride about London University as "the first and only medical school founded in England upon the full and extensive plan of the celebrated and systematic schools of the Continent and of Scotland"[51] suggests that this might have been one instance of the more general debate taking place between the radical faculty of London University and the conservative reformers, the practical London traditionalists of the hospital schools. Elliotson was not the only early faculty member touting London University's continental style. James Bennett, who had set up an English school in Paris, was hired initially as an anatomy demonstrator at London University. He advocated teaching anatomy in the style of the French, distinguishing tissue types and treating those tissues' general characters and functions.[52] Bennett was promoted to professor at London University when Granville Sharp Pattison, who was constantly maligned as old-fashioned, was removed from the position for incompetence in 1830.[53] Bennett, who was supposed to have been in a position of lower standing than Pattison, had long outshone Professor Pattison, even when still in his position as demonstrator, and was immensely popular with students.

With Bell and Pattison both gone by 1830, London University lost all representatives of the practical, anatomy-based approach to surgery, anatomy, and physiology that had been the hallmark of a London medical education since well before the founding of the University. Bell was angry enough to write in a rather vindictive way to his brother in 1831, "[t]he University, which was wont to be a subject of our correspondence, is going fast to the dogs; misrule and mismanagement are doing their work most efficiently."[54] For those men who resigned, men who hoped to reform medical education through the implementation of a practical curriculum at London University, the University had offered the possibility of uniting courses in medicine and surgery in order to provide a well-rounded education for the general practitioner, but such an education would be incomplete without a hospital. As Charles Bell described his position in his letter of resignation addressed to his students: the practice of medicine "is not sufficiently attended to by the rising generation of practitioners;—but it is an evil which the London University is incapable of remedying within the precincts of its present establishment."[55] Such an evil could not be remedied without a hospital, without the necessary material subject of medical science—sick bodies.

Historians have argued that London University's medical faculty was an important site in the battle by radicals to wrest control of medical education from the Tory elite, including the surgeons who controlled the Royal College of Surgeons and hospital appointments.[56] Read that way, the eventual incorporation of French morphology and experimental sciences into the curriculum, along with their attendant materialism and social radicalism, was simply one way of undermining the established hierarchy. But there were additional and competing reform agendas involved in the early years of London University, along with medical reformers whose politics were more local and whose aims had more to do with the professional practice of medicine than with imposing a major social and ideological shift. It is clear that both radical and conservative reformers acted for reasons that had little to do with large-scale politics: they left (or joined) London University in part because it was disorganized, in part because salaries were based on popularity with students, and in part because they had philosophical differences with their colleagues.

Although the radical medical men who sought to establish London University's curriculum on the pattern of medical science in France claimed to be the true medical scientists, their opponents also considered themselves to be systematizing medicine in order to establish medicine as a practical science as well—the two groups just disagreed about what might define a practical science. Bell and his cohort were decidedly antirevolutionary in their definition and aimed to reform the curriculum along lines that continued perceived British strengths. Bell and some of his loosely affiliated fellow opponents of "radical" experimental physiology built a well-organized and comprehensive educational program in medicine that was based on therapeutic practice. The juxtaposing of French science to London practice is something that these practitioners themselves adopted—the importation of French science would be done at the *expense of* practice. When London University became a radical institution and home to experimental sciences, it also lost its "British" character, and its faculty began to see it as cosmopolitan. But those reformers who meant to keep London medical education "British" and to achieve their own scientific reforms—by promoting practical experience, the integration of surgery and medicine into a single curriculum, and competition among faculty members and among institutions in order to reward teaching skills—went elsewhere. Often, like Bell, they went to the hospital schools that were built around practical medicine and its science.

HOSPITAL SCHOOLS: CREATING A PRACTICAL ALTERNATIVE TO RADICAL LONDON UNIVERSITY

While the faculty of London University continued to make a continental-style science of medicine by importing the latest life sciences (physiology, morphology, and comparative anatomy), Charles Bell joined the increasing number of conservative reformers located in hospital schools. Bell's reformers sought to create a medical science based on British anatomy and practical medicine, and they had found a new venue within which to further their pursuits. Growing directly out of the ward-walking practices of eighteenth-century London medical students,[57] nineteenth-century hospital medical schools essentially merged the institutional forms of private medical schools, universities, and hospital clinical lectures, crafting a form of medical education that was built around the practice of patient care. If London University came to represent the site of radical medicine in London and marked a change in the nature of medical education, then the hospital schools represented the prime site of conservative reform, accommodating the interests of the Royal Colleges by attempting reform through gradual shifts in education, with the needs of general practitioners in mind. The hospital medical schools, without the connections to laboratories and non-medical sciences, became a last bastion for the conservative reformers, but can also be seen as a product of and response to London University. Where hospital schools had previously offered ad hoc supplements to courses of anatomy in the private schools, by the 1830s they had become places of systematic education in clinical practice, partly in response to a model of systematic coursework set out by London University.

Hospital schools were established in two groupings—one set was established in the 1790s and the other set was born as a response to London University in the 1830s. In the 1790s, students could find schools and training opportunities associated with Guy's, St. Thomas's, St. Bartholomew's, and the London Hospital. In some cases, powerful private school lecturers managed to convince their hospitals to build lecture theaters to house some variety of courses, while in other cases private schools were located in close enough proximity to the hospitals that they became identified with those hospitals.[58] These schools were developed by those who improvised to meet a need. Several hospital schools—Westminster, St. George's, and the Middlesex (Bell's hospital)—by contrast developed as comprehensive medical schools later, in the 1830s and 1840s. They developed in a more self-conscious fashion and as an alternative to the medical training found at London University.

As natural as the emergence of hospital schools may seem, relationships between the governing boards of charitable hospitals, their staffs, and the teachers in their wards and schools were sometimes conflict-ridden. Hospital boards needed convincing that the creation of hospital medical schools would serve the hospitals' charitable mission—taking care of impoverished patients. These concerns are reflected in what appears to have been a standard sort of rhetoric used by hospital staff in appeals made to hospitals' governing boards for the establishment of schools. John Abernethy, a famous surgeon and lecturer at St. Bartholomew's Hospital, along with his colleagues, convinced the Board of Governors in 1787 to build an auditorium on hospital grounds, and later to build other facilities. The St. Bartholomew's Hospital staff made the argument that "it is impossible for the Medical Officers [. . .] to do all that is necessary to be done for the relief of Patients [without] recourse to the subordinate assistance of Students."[59] In addition to arguing that students were necessary to the normal functioning of the hospital, Abernethy and his colleagues also argued that medical education was important for furthering medicine.[60] Such appeals connected medical teaching to medical science and medical science to the curing of patients so as to make the support of medical teaching a necessary part of social charity. In order to get the support of the hospitals' usual benefactors, medical education in the hospital schools had to serve, or appear to serve, patients directly.

Over four decades later, Charles Bell used tactics similar to those of Abernethy in attempting to establish a hospital school at the Middlesex Hospital. After leaving London University in 1830, Bell focused much of his attention on finding or creating another faculty post for himself. In April 1835, six members of the Middlesex Hospital staff,[61] including Bell, presented an address to the hospital board that defined a hospital as a place "for the relief of those who are both sick and indigent" but also as a provider of "the grand means and materials of medical instruction";[62] the address ended with the claim that "there is an immediate connection between the promotion of its immediate purposes [treating the sick poor], and the extension of that science on which the relief and prevention of diseases depend."[63] The address noted that the only return that the house staff of the hospital received for their time and efforts came from the reputation they gained by working at the hospital and the compensation they received from students. That form of compensation was dwindling as other hospitals began to eclipse the Middlesex, offering students whole schools, with lectures in addition to opportunities to walk the wards.

Bell and his cosigners identified London University directly as a proximate cause for the need to establish a school, saying that the Middlesex used to get its hospital pupils from the private teachers on Windmill Street (including Bell and those who owned the school before and after him), but "upon the establishment of the London University and King's College [which was founded in 1829 as an Anglican alternative to London University], the school of associated teachers in Windmill-street was broken up," as both schools offered faculty positions to teachers who had formerly taught as independent, private instructors and offered students formalized instruction for which they could be credentialed. They also offered an example of how governors should deal with such a problem, citing those private school lecturers affiliated with St. George's Hospital as surgeons or doctors: "Those who belonged to St. George's transferred their lectures at once to the hospital [where they have been] granted ample accommodation for teaching the other branches of medical science within the building itself."[64]

Bell and his colleagues proposed to diminish the flow of students from the Middlesex to other hospitals by "establishing a complete school of medicine in avowed connection with the hospital, and under the sanction of its patrons."[65] To do so, they said, would promote the "efficient working of the charity, even in respect of its sick inmates, [which] should not be impaired by the want of a due supply of pupils, from whom must be chosen the house-surgeons, dressers and clinical assistants."[66] Their argument was that hospitals of the time drew their unpaid staff (house-surgeons, but also those who dressed wounds, etc.) from the ranks of their students. Those students who took on jobs like house-surgeon and dresser did some of the unglamorous work of the hospital in order to learn from the established staff members and to make a name for themselves, hoping one day to become hospital surgeons or physicians or well-paid private doctors. If those students went elsewhere for a more complete education, the hospital's patients would suffer for want of adequate personnel. Bell's subsequent opening address for the new school, printed in the *Lancet*, ended by repeating almost word for word the last line of the letter to the governors of the Middlesex, claiming an "ultimate connection between the promotion of [the hospital school's] immediate purposes, and the extension of that science on which relief and prevention of diseases depend."[67] These men tied the charitable mission of a hospital to medical education by asserting that making a hospital into a classroom would advance the clinical care of the hospital's patients, the incomes of the hospital staff, and the science of medicine.

In October 1835, when the new school opened, Bell became professor of surgery and anatomy there. He wrote to his brother on June 29, 1835, when the school had just been approved, "We have founded a school in the garden of the Middlesex Hospital. The building will be a complete little thing—theatre, museum, clinical class-room, and dissecting room." It was a school furnished with all the spaces necessary for a traditionalist's medical training—a theatre for dissection, museum for the study of pathological and anatomical specimens, and clinical classroom for training in therapeutics. There were no laboratories. It was a considerable undertaking for Bell, who wrote in that same letter, "you must admire my spirit to commence such an undertaking at this day. I promise to the extent of sixty lectures. To the work I have no objection, but there will be a great outlay also, though from the way in which it is taken up by our governors, I believe subscriptions will cover all expenses."[68] Such schools cost a great deal, both in terms of labor and potentially of one's own finances, but that was unsurprising to a man who had also paid to take over a private anatomy school in London, as

Bell had done with the Great Windmill Street School. Appeals to charitable boards were meant to obviate that need.

At first glance it appears that St. Bartholomew's and the Middlesex hospital schools were founded using similar rhetoric to overcome the concerns of a charitable board, but schools established in the late eighteenth century and those developed in the 1830s were born under very different circumstances. St. Bartholomew's Hospital School came into existence in 1788, more as a matter of convenience than design,[69] as a loose affiliation of instructors already teaching in private schools who thought that a centralized location would be convenient. Those hospital schools born in the 1830s, on the other hand, offered an alternative to London University rather than to the private schools, and thus aspired from the outset to be a "complete school of medicine."[70] All hospital schools, whether the loosely affiliated ones of the late eighteenth century or those developed to be complete medical schools in the nineteenth century, had to build the charitable mission of the hospital into their proposals, and that shaped the kind of educational program they promoted. The charitable nature of hospital care, however, was nicely suited to a conservative reformer like Bell, trading in the argument that British medicine was grounded in therapeutics and should be seen as superior to supposedly callous, French, materialistic, and unethical experimental medical science.

SYSTEMATIZING PRACTICAL MEDICINE: PEDAGOGICAL PROGRAMS WITHIN THE HOSPITAL SCHOOLS

As we have seen, during the early part of the nineteenth century the vast majority of practitioners of surgery and medicine, professions once considered to be distinct, dabbled in both. Thus, like London University, the hospital schools catered to general practitioners, attempting to reform medical training by providing students with a guide to general practice by offering courses that focused on routine diseases and injuries. In order to reflect the realities of practice, medicine and surgery were taught together. In his *Institutes of Surgery*, Charles Bell declared of the surgeon, "He is no longer a mere artist, a worker with his hands alone. The common sense of mankind has thrown into his department the treatment of many diseases, which require all the advantages of education hitherto imparted to the physician. The studies of the physician and of the surgeon have become the same."[71]

These general practitioners were served by two new sets of institutions in the British capital. While London University was planned and built to satisfy the need for professional (not only medical) education in London, London's hospital schools grew up in a more unplanned manner, as hospital practitioners, mostly MDs and surgeons, banded together to offer full curricula in a single, convenient location, sometimes absorbing the small private schools that surrounded them. By 1830, they were being developed systematically, as the practical response of conservative reformers to London University. The hospital schools had to be made to serve charitable ends in order to gain the support of the hospitals' governing boards. The hospitals' staffs made their appeals to the governors by claiming that hospital schools advanced practical medical science (therapeutics) and that schools offered free sources of labor to the hospital. These appeals ultimately contributed to the way that British practitioners saw themselves and their medicine.

While London's hospital schools were founded during the same period as that in which France's famous clinics were born,[72] the two sets of clinical schools, according to the British, had different origins and strengths. That the clinical training in London took place in charitable hospitals mattered: autopsies were uncommon, patients were not subjected to repeated physical examinations, and surgical operations were less frequently performed for a crowd, all of which distinguished it from French clinical training in the same period.[73] These differences allowed British medical practitioners to depict education in their country as being conducted by men who were pragmatic and humane, as opposed to the scientifically minded but unfeeling medical scientists abroad, who devoted themselves to vivisection and pathology without regard for preserving life.[74] While sciences that existed independently of medical practice, like comparative anatomy, physiology, and pathology, became the markers of medical science for radical reformers and Francophiles, British hospital school teachers and conservative reformers also saw their "British" educational programs as scientific—the scientific element derived from a well-designed system, or progression and interrelationship, of courses. And, more importantly to the hospital practitioners hoping to reform medicine in ways that conserved what was best about British medicine, it was scientific and practical in a way that London University could never be (no matter how many new sciences it imported from the Continent), so long as it did not have a hospital.

In 1838, Benjamin Brodie, surgeon at St. George's Hospital (the other hospital that, along with the Middlesex, was originally to be affiliated with

London University), wrote in his *Introductory Discourse on the Studies Required for the Medical Profession*: "[W]hile engaged in attendance on the hospital, always bear in mind that there is no one of your other studies which, as to real importance, can compete with this."[75] Hospital schools possessed obvious advantages in practical training over any other venue. In addition to offering clinical training, the new hospital schools of the 1830s integrated classroom lectures on the medical sciences into their curriculum. Their courses outside the hospital wards, however, tended to be limited to subjects with comparable relevance to practice.

The Middlesex Hospital was a fairly typical hospital school, offering courses in medicine, surgery, anatomy and physiology, midwifery, therapeutics, chemistry, forensic medicine, and botany, all of which were intended to be practical in nature.[76] (It is notable that the Middlesex did not separate its physiology course from that on anatomy, or offer a separate course on comparative anatomy as did London University.)

Nineteenth-century hospital schools presented a space in which education in the practice of medicine could be systematized in a way that stressed therapeutics. The joint emphasis on the practical and on its philosophical or principled underpinnings was very much a part of the program of a conservative reformer like Bell, and followed directly from complaints about London University's deficiencies—its lack of a hospital, the site of instruction in the practice of medicine. Charles Bell explained how the structure of the hospital served his pedagogical purpose in his "Clinical Lecture on Diseases of the Spine," delivered in 1835 at the Middlesex Hospital School:

> I have requested you to come into the theatre rather than your [class]room in the hospital, that I might show you these preparations in connexion with the cases you have seen. And now observe the advantage of giving clinical remarks to those [to] whom I know the right elements have been taught, instead of addressing gentlemen from three or four neighboring schools, who have in all likelihood no ideas in common with me. I will furnish you with an example of how easy it is to give such lectures to those who have been initiated in the principles.[77]

Bell stressed the importance of principles and of experience jointly—principles were the things that allowed a student to make sense of whatever he observed; they constituted the natural philosophy that allowed one, for example, to assert that form and function should be related. He went on, then, to jump straight into a case that illustrated those principles in practice:

> If you go into Percy ward you will there find a man lying with a wandering and bewildered eye, with a very very pale face, and spasmodic twitching of the eye ball . . . This man has fallen upon the vertex of his head; and were I to enter upon the consideration of the case with those who had not gone through the demonstration of the bones of the cranium with us, you know full well that I should be obliged to enter upon the whole structure of the skull, and the mechanical principles on which it is built. But now, with one word, one half sentence, I can say to you there is the example of which I have been speaking. Here a blow has been inflicted upon the upper part of the parietal bone, and you see the effect upon the temporal bone, and upon the ear . . . It is not requisite that I should go into the whole proof, and repeat the demonstrations; I have merely to say, that this is an illustration of the principles I formerly established.[78]

In other words, Bell considered it crucial when in the lecture theater to be able to make reference to hospital cases to which all of his students had access and, equally importantly, to know that his students had the same background in anatomy and other basic medical sciences when explaining practical or clinical medicine in the hospital. These interrelated aspects of medical education—experience related to practice and underlying principles—necessitated a *system* of medical education, a thorough training not just through apprenticeship, but one that united a study of the principles of human anatomy and disease with that of therapeutic practice. Bell drew on clinical cases from the hospital in order to illustrate physiological and pathological principles.[79] The development of hospital schools advocated by medical men like Bell had to do with establishing a pedagogical system in which basic sciences like anatomy were taught systematically, with reference to each other and to actual practice in the hospital wards.

In his introductory address at the Middlesex Hospital School, Bell directly juxtaposed the new hospital school to his old affiliation, London University. In the *Lancet*'s report of Bell's address: the Middlesex "had been established in order to counteract the effect of a rival party [i.e., London University], who had deprived of its pupils the Middlesex Hospital, the governors of which were no sooner fully informed of the fact, than they enthusiastically came forward and supplied the funds necessary to institute a medical school in connexion with the hospital."[80] Thus, Bell seems to have made it clear in the address that there would be substantial differences between London University and the Middlesex Hospital School. He said that he had thought it bad for any hospital school to have both a professor and a demonstrator of anatomy (as London University did), since the pupils, the *Lancet* reported Bell as saying, "would be more intimate with the demonstrator than the professor [. . .]. Anatomy was not to be learned without the constant presence of the teacher in the dissecting-room, and he thought that the proper plan was for the teacher himself, the 'professor' or the 'demonstrator,' whichever name they chose to give to the *teacher*, should put on the sleeves and apron, and demonstrate in the dissecting room, as he (Sir Charles Bell) had done."[81] The *science* of practical medicine that Bell hoped to establish would unite the practical and philosophical elements of anatomy, rather than viewing the practical parts of anatomy instruction as derivative. The teacher-demonstrator would need to be in the room with students, who were literally practicing surgery and medicine through dissection. This important shift shows the commitment of Bell and his peers to teaching all of medicine's component sciences as practical (both in the sense that they were not theoretical and in the sense that they had a relationship to therapeutic practice). For Bell, the Middlesex Hospital School represented the realization of his ambitions. He said of its establishment, in an 1835 letter to his brother, that "it has ever been my pride to join the pursuits of science (and lecturing is of all conditions the most conducive to scientific pursuits) and practice. . . . It has been in truth under that conviction that I have just formed a School at the Middlesex Hospital and you cannot conceive a prettier thing than that School."[82]

Anatomy was taught in the hospital schools as the fundamental medical science, one that informed surgery and medical therapeutics. Physiology, that experimental French science, was taught as part of the anatomy course at St. Bartholomew's until the 1840s, on the view that physiology was subservient to anatomy.[83] And in 1838, Brodie, a friendly acquaintance of Bell's, surgeon and teacher at St. George's Hospital, and fellow conservative reformer, described the way that the sciences were divided at St. George's: "Anatomy and Physiology are one science, and to teach them separately is about as absurd as it would be to divide Astronomy into two sciences, the one teaching the figure and size of the heavenly bodies, and the other their motions."[84] This sort of linking of anatomy and physiology, seen in the course registers of St. Bartholomew's, the Middlesex, and St. George's, and with anatomy being granted primacy, was typical of the hospital schools, again placing them in sharp opposition to London University, with its emphasis on new experimental and laboratory life sciences. Brodie explicitly contrasted London's practical courses on anatomy to those of the French:

> I have no doubt that the praises which are bestowed on some of the continental anatomists are well founded: that there are universities in which the anatomi-

> cal professors, devoting their whole time, and industry, and intellect, to this one pursuit, explain the mysteries of minute anatomy at greater length and with more precision, than the teachers here: but nevertheless, I assert that ours is the better method with a view to the education of those who wish to become not mere philosophers, but skilful and useful practitioners.[85]

It was, according to men like Brodie and Bell, the superior British style of anatomy, which was taught with reference to *practice*, that lay at the core of the medical science taught in British hospital schools.

Physiology in the hospital schools was accorded a lower place than anatomy partly because, in its British form, it was seen as a derivative discipline, based on anatomy and not on vivisection or another independent set of methods. Charles Bell wrote of his anatomy course in 1838: "The objects which should occupy the young surgeon in the dissecting-room are these: Every thing done should have reference to the living body—the forces which act on the bones and ligaments—the classification of the muscles, and their action in cases of fracture and dislocation."[86] In other words, anatomy was meant to encompass physiology. Physiology was also given low standing because on its own it was seen as having very little therapeutic value. Other specialties, like pathology and materia medica, covered treatment of a diseased body, whereas physiology simply explained the workings of a healthy body without reference to therapeutics. Although "practical science" seemed to be the ambition of all medical reformers, for Bell and his cohort, the word "practical" carried with it connotations of therapeutic relevance that were not critical to the sort of "practical science" defined and constructed by the radical, continental-style reformers of London University.

Brodie described pathology at St. George's Hospital School by likening it to physiology, in that it relied heavily on other sciences: "In like manner, Pathology is not taught here as a separate science, but you receive your instructions in it from the Lecturers on the practice of physic and surgery, who, while they explain the changes of function or structure, which constitute disease, point out also the symptoms by which the existence of these changes is indicated in the living body, and the means to be employed for the patients' relief."[87] The sort of disease history of a patient in the hospital given by Bell in his "Clinical Lecture on Diseases of the Spine," with its reference to anatomy and pathological structures, was exactly the sort of pathology that Brodie, Bell, and the conservative reformers of the hospital schools thought should be taught. Everything should be done with reference to therapeutic practice. Brodie explained further the benefits of such a system: "while you are taught Pathology, you are taught also its uses and application; and these different subjects, brought under your view at the same time, serve mutually to elucidate each other; for, while Pathology assists you in obtaining a knowledge of symptoms, the study of symptoms, and of the operation of remedies, contributes in no small degree to extend your knowledge of Pathology."[88] To such men, the kind of pathology practiced in the laboratory with dead tissues, a kind associated with London University and the French, was an uninformed pathology and one with very limited applicability, as it was most useful only in highly localized diseases.[89] In all other cases, pathology had to be studied in the hospital as much as the dissecting room, combining anatomy, physiology, and detailed case histories.

The stress on practical therapeutics in the development of a medical curriculum, and even in the defining of medical sciences like physiology and pathology, distinguished the hospital schools from both London University and continental-style medical schools (including the University of Edinburgh). Unlike universities, which had interests in, and space for, the pursuit of knowledge unrelated to patient care, hospital officers' chief emphasis had to be on the treatment of patients. As a site for the development of medical science and medical education, therefore, hospital schools' pedagogical programs were fundamentally shaped by the joining of practice and charitable alleviation of suffering, both of which can be seen in the rhetoric of British hospital teachers. This coupling appears in Charles Bell's 1838 textbook *Institutes of Surgery*, in which he wrote that "moderation" was key: "[t]he student sees there great operations dexterously performed amidst the applause of hundreds: But it would be well for him to study the consequences of these exhibitions;—to follow the patient into the ward, there to learn the difference between dissecting and operating;—to see how much the human constitution can bear, and be directed to the stuff of the powers of life and of the constitution."[90] The suffering of the patient and attention to the human constitution here distinguish medical training in a hospital, on living patients, from the type of surgery that could be practiced on a cadaver, or pathology that could be learned from autopsies. Such a perspective is also evident in the advice of Benjamin Brodie in his address to students from the same year: "never losing sight of the ultimate object of all your investigations, you must estimate the value of whatever other knowledge you acquire by the degree in which you find it to be directly or indirectly applicable to the healing art."[91] This focus on the sick in the hospital, on their care over the long term, illustrates the utility of the hospital as a place to teach humility, and to highlight the alleviation of suffering rather than technical operative success as the marker of achievement in medicine. It was this focus on the patient that the British saw as distinguishing them from the increasingly influential French practitioners, and it was this character that was built into the charitable hospital's school.

CONCLUSION

In the late 1820s and early 1830s, London University and London's hospital schools offered new institutions through which medical education could be reformed. London University became the home to radical reformers who used the new institution to import sciences like morphology and experimental physiology. Charles Bell and other conservative reformers strove to create more practical training for surgeons, physicians, and general practitioners. According to them, London's main advantage as a center of teaching was its many charitable hospitals, its many locations at which to *practice* medicine. Bell's first university home, London University, was conceived as an attempt to systematize medical education, and that mission was brought to the hospital schools as the medical community and even the community within London University became divided about just *how* that reform should work in practice. Hospital schools like the Middlesex were developed in the 1830s to offer a comprehensive, practical alternative to London University. In both cases, the institutions were messy, imperfect instantiations of the philosophies that they were built to represent.

In 1835, when the University had the funds to establish a hospital of its own and on its own terms, the building of London University's University College Hospital was begun outside the control of the Royal College of Surgeons and with careful consideration of the possibility of the hospital becoming more important than the teaching facilities, of difficulties over running costs, and of access to patients.[92] By this time, James Bennet, Robert Grant, and others had helped to make London University a "French school in England [. . .] with its insistence on prestigious European science before local London practice,"[93] as Adrian Desmond has characterized it. But those who prioritized "local London practice" had an alternative set of ambitions for medical science and were not simply resisting change and improvement. By the 1830s, the venue for their reforms and their model of a practical medical science was the hospital school, a kind of pedagogical institution that they spoke of as being entirely British in its style and ambitions.

By the late 1830s, the universities of London, London University and

King's College, had formed partnerships with the existing hospital schools. These two institutional forms, the university and the hospital school, had become symbiotic: the London Hospital School (another hospital school that predated the founding of London University), the Middlesex Hospital School, St. George's Hospital School, and St. Bartholomew's Hospital School had all become loosely affiliated with London University. Each institution operated independently, with the University having no power to inspect or control the hospital schools, but the hospital schools' students were able to take advantage of the University's laboratory classes, while the University's students were able to pursue clinical experience at the hospital schools.[94] Students from outside London also began to do classroom training at another university (Oxford or Cambridge, for instance) and hospital training at one of the city's hospitals, disrupting the London-based school of medicine that had previously existed. This sort of compromise actually represented something of a victory for London University, which had driven out competition from separate, independently functioning hospital schools while acquiring additional hospitals on its own terms. The educational reformers of the hospital schools, on the other hand, had intended their pedagogical program to be different from London University's and to be overwhelmingly practical in nature. The increasing incorporation of laboratory classes and classes without direct reference to medical practice into the standard medical curriculum would slowly dilute the practical British character of the medical science being built in the classrooms of the hospital schools.

Historians have often remarked on the disconnectedness of medical science from medical practice in the nineteenth century, assuming that science was simply used as a rhetorical tool and a legitimizing bulwark for medicine.[95] But early attempts to make a science out of medicine in Britain were made by men whose focus was actually on medical practice itself—the attempt really to systematize medical education and to ground it in practice was one that the actors themselves termed an attempt to create a "science of medicine"; as Bell wrote, "it has ever been my pride to join the pursuits of science . . . with practice."[96] These men wanted to unite surgery and medicine, as well as practice and theory, thereby to elevate the practical art of surgery while making a system of medical learning, one that would use hospitals as "living museums of disease" and that would rely on careful deductions from anatomical dissection to create an experience-based physiology.[97] Doing so would help to fashion a particularly British style of medical training in practical therapeutics meant to aid the general practitioner. Experimentation and laboratories were not the exclusive markers of scientific medicine: the origins of medical science can also be seen taking shape in the classrooms of the early nineteenth century. As Charles Bell declared in his Inaugural Lecture at the University of London, "I deam [*sic*] the right teaching in any department of science the surest way of improving it."[98]

CHAPTER FIVE

Defining a Discovery: Changes in British Medical Culture and the Priority Dispute over the Discovery of the Roots of Motor and Sensory Nerves

> I took the opportunity of saying that in stating the course of my own reflections, and the mode in which I had proceeded, I left controversy to my younger friends; that in spending the greater part of my life in the duty of teaching, I had educated many to the profession who knew the correctness of my statements, the mode and succession in which my ideas had developed themselves, and who were as willing as capable of defending me against illiberal attacks.
>
> CHARLES BELL, in a note found among his papers, written upon his return to Edinburgh for the British Association for the Advancement of Science meeting in 1834[1]

In 1836, Charles Bell accepted a professorship from the University of Edinburgh. George wrote to him saying: "You are now an adopted member of this University, and with the unanimous assent and acclamation of a Town Council composed of persons of all parties, chosen by the several wards of this intellectual City. And surely never was an offer more honourable to an individual, for I do assure you it has the approbation of all ranks and classes of men, and of none more than the Professors of the University."[2] Bell had come full circle. He left Edinburgh in 1804 because his brother John had made enemies at the university and among the medical profession, leaving Bell few options in the town. By 1836, Charles had sufficiently demonstrated his skill as a surgeon and anatomist, but equally importantly, he had also proven himself agreeable and gentlemanly, had cultivated the right spectrum of social and professional patrons, such that he had won the approval of "persons of all parties," "all ranks and classes of men." He was taken back by the city that he had left and to the university that had driven him out.

In Edinburgh, the striving life that Bell had led in London quieted a bit. He went fishing and enjoyed nature, often taking his sketchbook. Edinburgh seemed the right place for Bell to finish out his career. He wrote to John Richardson, a companion in fishing:

> London is well for the young and expectant, or the dull old citizen who has passed the vigour of life on a stool, and who continues to play with his thousands as a young coxcomb, whom Nature meant for a fool, lays his bets on the balls of a billiard-table. No, it won't do for us. We must waken with the rustling leaves at the window, and the fresh breeze off the woods and garden, and we must get learned in the weather, and watch the clouds and the water, while we have eyes to tie a knot and choose a fly.

In addition to the pleasures of nature, Bell also enjoyed the company of old friends and of his brother, writing about his various social gatherings: "My Lord Cockburn I meet, who is still 'Cockie.' The Jeffreys and Listers keep my wife in play. We dine with them at Ratho to-morrow, with Jeffrey and Sir J. Dalrymple on Tuesday, and next day we (the College of Surgeons) give a public dinner to Sir Astley Cooper; indeed, he and Lady Cooper being here has *taigled* [tickled] me very much."[3] Bell's Scotticisms finally flourished unchallenged. Back with his brother George, as he had yearned to be when first in London, Bell's letters from that period are almost all to Richardson, his closest friend in London.

Bell remained a committed teacher in Edinburgh. He wrote to his old friend Henry Brougham about his ambitions to renew some of what he found wanting there, saying, "I hope to do much for this university in prevailing with my colleagues to do some things which have fallen into neglect—e.g. anatomical rooms, museum of anatomy, examinations in the classroom [all] require renewed attention."[4] His concern to improve the anatomical rooms and museum suggest that even at the end of his life, Bell's commitment to the teaching philosophy described in chapter 2 of this book remained unchanged. Anatomy and teaching from museum objects remained at the heart of an encompassing educational system that also incorporated traditional subjects. Bell wrote of Latin examinations that he conducted, "as I hate all false pretences, I every morning read diligently at my old school books; and commencing in duty, I proceed with a quiet pleasure I have not known for many a day. It is an unambitious dreamy use of time that gives perfect rest."[5] Delving back into his old books, he found peace, and in his position of respectability, he found comfort. "Men waste their lives from want of method," he wrote to John Richardson in 1839, "I am convinced that I am doing good to mankind (don't laugh at me) by my lectures, and on the whole, tho' fatiguing, they are pleasant to me."[6]

Still, Bell found himself without the financial means he had expected, a consistent problem throughout his life. Bell wrote to Henry Brougham of his position at Edinburgh, "[m]y only disappointment is that there is no salary which in all justice there ought to be and which I understood there was."[7] His health worsened. His stomach was a continual source of grief. Of it, Bell wrote, "My stomach is still my great evil, and makes me unhappy and full of forebodings. Exercise and starving, I am confident, are my only cure."[8] During his time in Edinburgh, in the last years of his life, Bell and his most dedicated students worked to preserve his standing in a changing medical world. Alexander Shaw, still in London, published his defense of Bell's work on the nervous system, *Narrative of the Discoveries of Sir Charles Bell in the Nervous System*, in 1839.[9] It was intended to set right a dispute over priority that had plagued Bell through much of his career. That dispute is the subject of this chapter, though I am in no way attempting to arbitrate its merits or to evaluate Bell's place in the history of neuroscience. Instead, the dispute highlights the ways in which the medical world was changing around Bell as he moved through it and, more generally, reveals the ways in which priority disputes can change shape over time and therefore should not be condensed into two distinct and flattened arguments.

ORIGINS OF THE DISCOVERY

The dispute, or at least the discovery that started it, had begun in 1811, when Charles Bell was trying to make his living in the crowded London medical marketplace.[10] He had a little book, *Idea of a New Anatomy of the Brain*, printed for distribution to his friends and other natural philosophers and medical men.[11] The book contained what Bell considered to be a great discovery on the workings of the nerves and brain: "considering that the spinal nerves have a double root, and being of opinion that the properties of the nerves are derived from their connexions with the parts of the brain, I thought that I had an opportunity of putting my opinion to the test of experiments, and of proving, at the same time, that nerves of different endowments were in the same chord [*sic*], and held together by the same sheath."[12] The experiments that Bell mentions so casually were both technically difficult and hard for Bell to stomach, according to his own accounts in personal correspondence. In the text, however, they are presented as unremarkable. Bell did not apologize for them or emphasize the difficulty of the surgery required to conduct them. Bell's emphasis was clearly on the logic of the system of the nerves that he had constructed and not on its experimental proof.[13] It is also evident from the list of the book's recipients—scientists, doctors, and wealthy or powerful men who might provide patronage—that Bell intended this pamphlet for a small, professional or else highly educated audience.[14]

By 1823, when the Frenchman François Magendie claimed a very similar discovery for himself, the medical world had changed a great deal. Journals were beginning to proliferate, and Magendie, who contested Bell's claims to discovery on the very basis of their inadequate publication, presented his own work in a journal he had himself founded.[15] Vivisection, which became a part of the debate between Bell and Magendie, had become a familiar and much reviled practice in Britain and was the subject of legislation there in 1825.[16] When speaking before the House of Commons that year to promote his anticruelty legislation, MP Richard Martin adduced the callousness of vivisectionists, using Magendie as his prime example. Hansard reported Martin's address:

> In the course of the last year this man, at one of his anatomical theatres, exhibited a series of experiments so atrocious as almost to shock belief. This M. Magendie got a lady's greyhound. First of all he nailed its front, and then its hind, paws with the bluntest spikes that he could find, giving as reason that the poor beast, in its agony, might tear away from the spikes if they were at all sharp or cutting. He then doubled up its long ears, and nailed them down with similar spikes. (Cries of "Shame!") He then made a gash down the middle of the face, and proceeded to dissect all the nerves on one side of it. . . . After he had finished these operations, this surgical butcher then turned to the spectators, and said: "I have now finished my operations on one side of this dog's head, and I shall reserve the other side till to-morrow. If the servant takes care of him for the night, I am of the opinion that I shall be able to continue my operations upon him to-morrow with as much satisfaction to us all as I have done today; but if not, ALTHOUGH HE MAY HAVE LOST THE VIVACITY HE HAS SHOWN TO-DAY, I shall have the opportunity of cutting him up alive, and showing you the motion of the heart."[17]

The story is probably fictitious,[18] but it clearly served its inflammatory purpose. After telling it, Martin made a point of emphasizing that John Abernethy and Everard Home, two well-known *British* surgeon-anatomists and professional patrons of Bell's, had along with other medical men written statements condemning such cruelty. Vivisection became an element of the Bell-Magendie priority dispute that resonated with the professional and nationalistic politics and rhetoric of the British medical community, as discussed in the previous two chapters.[19] Vivisection became a significant element of what turned into, in part, a nationalistic priority dispute, but it was never a straightforward part, and it exemplifies the strategic positioning, defining, and redefining that occurred over the life of Bell's dispute with Magendie.

The story of the discovery reveals much about the way in which the British medical community of the early nineteenth century was structured and the way in which it was changing. Credit for *a* discovery regarding the roots of the nerves was claimed by many, and those would-be discoverers presented overlapping, but not identical, definitions of their discovery; an agreement over what was actually discovered (though not who discovered it) emerged out of conflict, negotiation, and revision. The process whereby the definitions of, and credit for, the discovery were crafted is the subject of this chapter. Because the discovery was contested, first by the Frenchman François Magendie and later by Bell's own student and countryman Herbert Mayo, it offers an opportunity to evaluate the ways in which British medical practitioners and physiologists defined themselves over time in relation to the French and in relation to each other. Allegiances to fellow countrymen and to particular medical factions helped to create a complicated set of shifting priorities that developed through, and were revealed by, responses to Bell's work within the fragmented British medical community.

Though Bell received much acclaim from his contemporaries for his work on the nerves,[20] his work was also criticized by many and changed shape often. It was continually being defined and redefined by Bell, his student-advocates, his defenders, and his detractors. The complicated disputes surrounding the substance of the discovery of the roots of motor and sensory spinal nerves, as well as the changing definitions of discovery never really acknowledged by the historical actors themselves, provide an opportunity to glimpse the ways in which professional communities were structured and the ways in which their members used various forms of print and other tools of communication to build alliances and assert priority claims.[21]

This chapter begins by returning to the world of an ambitious young London anatomist in 1810 and 1811. Bell's "discovery" began, as chapter 1 discusses, as a self-fashioned work of greatness. On May 21, 1807, he wrote to his brother, "I am casting about for a subject to make something new of. I have been thinking about the brain—of mind—of madness." And later, on November 31, 1807, Bell wrote, "My surgical books and lectures you will soon see eclipsed by my character as an anatomist and physiologist. I really think this new view of the Anatomy of the Brain will strike more than the discovery of lymphatics being absorbents."[22] He had only three students and little income at the time, and this emphasis on fame and discovery was, as we have seen, partly an attempt to secure financial stability. But Bell always coveted a place among the natural philosophers; he once asserted, "[I]f I am to be anything, it is from connection with Natural Philosophy by Anatomy."[23] He therefore continued to assert the significance of his discovery long after he had established himself. In 1821, he wrote: "I know better than others can tell me what is to become of this. It gives me a power of doing what I choose now, and will hereafter put me beside Harvey—but this is in your ear. Harvey was said to have had the way prepared for him so that he could not miss it—so fools argue. But the discoverer of the nervous system had nobody to go before him. The discoveries of anatomists had only made the matter more intricate and abstruse."[24] That passage reveals well what Bell thought his contribution to this much-studied subject was—he had taken a muddle of work by previous anatomists and made it clear and simple (figure 13).

As is evident from the title "Anatomy of the Brain," Bell did not begin his work with the anatomy of the nerves as his subject. He began his work on the nerves in an attempt to understand the functions of the brain (figure 14).[25] According to Bell, previous anatomists had assumed that the brain acted as a single sensorium, that the whole thing together interpreted sensations; they did not break the brain down into anatomically and physiologically distinct sections. In such a view, nerves were each fitted to receive a particular impression from their environment and differed only in degrees of sensibility, while individual nerves carried both impressions to the brain and the force of the will from it.[26] Bell attempted to map out the functions of the brain by mapping out its anatomy, adopting both methodologies and objectives well known among British anatomists. Like phrenologists in Edinburgh[27] and those who worked on natural theology,[28] Bell believed in the fundamental principle that form was revealing of function. He saw the body in *Idea of a New Anatomy* as "a system great beyond our imperfect comprehension, formed as it should seem at once in wisdom; not pieced together like the work of human ingenuity," and thus as an intelligible system and one whose workings could be revealed by investigating the body's structures.[29] He saw his own work, therefore, as becoming (if not yet being) a kind of anatomical study, coupled with pathology to reveal physiology, that

FIGURE 13. Engraving "Nerves of the Neck" drawn by Charles Bell, in his *A Series of Engravings Explaining the Course of the Nerves* (1803), plate ii. Bell wrote in the text accompanying the engravings that the nerves of the head, neck, and chest might seem to a student to possess "an intricacy of connexions, which would require the labour of many months to disentangle," and that in viewing that complexity, "[h]e may be inclined to think this apparent confusion accidental." But Bell says that evident difficulty should not discourage the student, for, as he himself set out to show, "there is here perfect order! That each little knot or twig of Nerve, visible in one body, is seen also in all the others. That the parts are infinitely minute, but that intricacy or confusion are words not applicable to Animal Structure . . ." (preface, p. 5). The beauty of order in man's anatomy is reflected in the beauty of Bell's images—the face of the dissected man depicted is detailed and individualized, with a sense of peace and dignity. Image courtesy of the Wellcome Library, London.

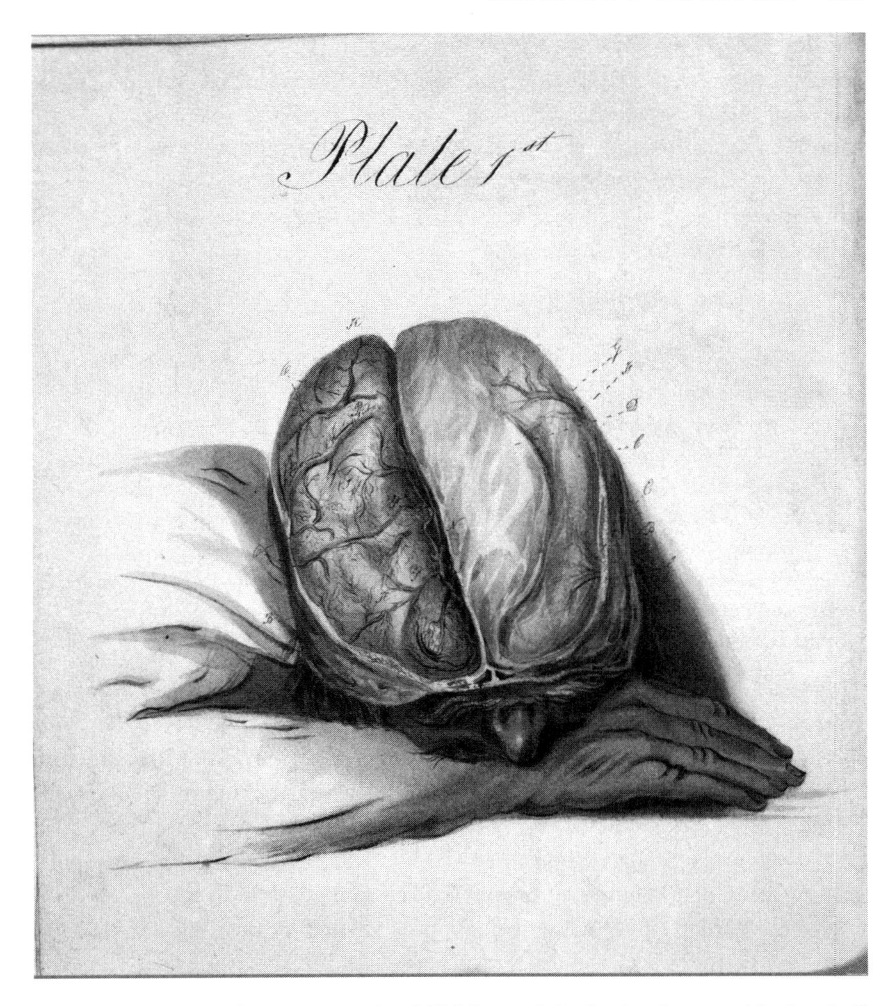

FIGURE 14. General anatomy and subdivisions of the brain, drawn with the skull cap removed, originally drawn by Charles Bell for Charles Bell, *Anatomy of the Brain* (London: Longman, Hurst, Rees, and Orme, 1802), plate 1. This 1823 manuscript version is hand colored after the original drawings. Image courtesy of the Wellcome Library, London.

could replace vague notions about the functions of organs as a whole with descriptions of specific processes carried out by specific anatomical systems among those organs. Having found it difficult to work on the brain directly, Bell traced the brain outward through the nerves, asserting that the cerebrum and cerebellum were "different in function as in form" and that the bundled nerves that he traced from the brain into the body were "distinct in office as they are in origin from the brain."[30] In other words, by looking at the anatomy of the nerves specifically, Bell thought he could map out their functions as well as those of the parts of the brain itself.

In 1810, Charles Bell first described in a detailed fashion his progress on the discovery he had set out to make. He detailed experiments on the spinal nerves in a letter to his brother:

> It occurred to me that there were four grand divisions of the brain, so were there four grand divisions of the spinal marrow; first, a lateral division, then a division into the back and fore-part. Next it occurred to me that all the spinal nerves had within the sheath of the spinal marrow two roots—one from the back part, another from before. Whenever this occurred to me I thought that I had obtained a method of inquiry into the function of the parts of the brain.
>
> Experiment 1. I opened the spine and pricked and injured the *posterior* filaments of the nerves—no motion of the muscles followed. I then touched the *anterior* division—immediately the parts were convulsed.
>
> Experiment 2. I now destroyed the posterior part of *the spinal marrow* by the point of a needle—no convulsive movement followed. I injured the anterior part and the animal was convulsed.[31]

This letter seems to describe in narrative fashion the development of an idea from anatomy and then the simple test of that idea through vivisection, both of which were carried out in Bell's classrooms, with the aid of his students. The idea, its presentation and ordering, and the experiment involved all formed important elements of the priority dispute that ensued.

It was through his teaching that Bell first publicized his work and gained some of his fiercest allies and advocates for his discovery. *An Idea of a New Anatomy of the Brain*, privately circulated in 1811, gives a sense of how Bell conceived of his discovery before he produced a grand folio for general publication, and before that discovery was contested. Bell opened the text by saying that anatomists did not understand the brain and nerves, that notions about the nervous system were vague and based in analogy rather than precise description.[32] His friends and colleagues, he said, misunderstood him and his aims in investigating the brain, assuming that he was looking for the seat of the soul, often thought to be in the brain.[33] Such an ambition, Bell claimed, would have been presumptuous and foolish. Instead, his purpose was to investigate the brain as anatomists and physiologists would any other organ: through detailed anatomical dissections and deductions from those dissections. Setting up his own work against those of distinguished predecessors, Bell claimed "to offer reasons for believing, that the cerebrum and cerebellum are different in function as in form; that the parts of the cerebrum have different functions; and that the nerves which we trace in the body are not single nerves possessing various powers, but bundles of nerves, whose filaments are united for the convenience of distribution but which are distinct in office as they are in origin, from the brain."[34] Bell's work, he believed, had the added virtue of simplicity, an advantage that he would rely on in defense of his contribution throughout his lifetime.

In the introduction to his *Idea of a New Anatomy*, Bell asserted that before him, the more one knew about previous work on the anatomy of the brain, the more confused one became. A hodgepodge of previously established facts and theories, many of which contradicted each other, made the brain like a maze. In great part because of his faith in natural theology, Bell found such confusion unacceptable; he placed a high premium on the elegance of systems of anatomical parts working together, describing the nerves as such a system in his text. He wrote to his brother, "I establish thus a kind of circulation, as it were. In this inquiry I describe many new connections. The whole opens up in a new and simple light."[35] Here again Bell used Harvey's discovery of blood circulation as a prototype. It is noteworthy that Bell almost always compared his work to Harvey or Hunter when aiming to establish its significance—his own self-presentation very much accorded with the rhetoric of conservative reform, invoking British tradition in discussions of future greatness. In addition to hoping that his own work would be as monumental as that of his British predecessor, Bell appreciated the elegance of Harvey's anatomical system. The very notion of circulation, in which one type of anatomical part carried something vital from the center to the periphery, and a parallel and similar part carried things back to the center, seemed to Bell to apply to the nerves.

Bell went on to detail his findings on the brain, the cerebellum, and the double roots of the spinal nerves. He began by describing the anatomy of the brain itself, noting the ways in which the cerebrum and cerebellum were clearly anatomically distinct—in form, in color, and in vascular structure—and remained distinct in various species of the animal kingdom. Bell found that the cerebrum, home of sensory perception and thought, varied in size proportional to the sophistication of the species and its nervous system. He then matter-of-factly described an experiment he performed on a living donkey to confirm the functions of the parts of the brain: "On laying bare the roots of the spinal nerves, I found that I could cut across the posterior fasciculus of nerves, which took its origin from the posterior portion of the spinal marrow without convulsing the muscles of the back; but that in touching the anterior fasciculus with the point of the knife, the muscles of the back were immediately convulsed."[36] It is on the basis of this experiment that Bell claimed to have concluded that every nerve with a double function must have a double root, a connection with both the cerebrum and the cerebellum. In his *Idea of a New Anatomy*, Bell called these double functions "sensible" and "insensible," and later, "nerves of sense" and "nerves of motion." The cerebrum, he claimed, united the body with the world, containing the nerves bearing sensory impressions from the outside world and the nerves that carry the force of will to the various parts of the body, while the cerebellum handled the nerves responsible for basic functions of the body itself. Bell ended the piece by summarizing the functions of the nerves, saying, "[t]hrough the nerves of sense, the sensorium receives impressions, but the will is expressed through the medium of the nerves of motion."[37] The book is short and laid out in a sketchy fashion, but it is clear that Bell thought the work was significant.

An understanding of the brain and of the nerves had implications beyond physiology and medicine as well. Investigations of the brain and nerves could appear to lead to materialism or sacrilege, and they could also have political and social overtones, as was the case with phrenology.[38] Here, by talking about the will, Bell is using a term that he surely knew had a philosophical history (for his letters suggest that he read philosophers like John Locke): the exercise of free will was an important component of debates about materialism, about religion, and about democracy in eighteenth- and early nineteenth-century Britain.[39]

Bell's method of pursuing his ideas about the brain and nerves became a central part of later priority disputes and is worth examining in more detail, as it is also representative of a British style of anatomico-physiology that was increasingly the subject of debate. As I have described above, Bell surmised that each root's properties related to the part of the brain with which it connected and that nerves with different functions were bundled together into a single cord as they made their way out into the body. In order to put such ideas to the test, Bell developed two experiments, described only briefly in *Idea of a New Anatomy* in the fashion depicted above—one in which he opened the spine and "pricked and injured the posterior filaments of the nerves . . . then touched the anterior division" and one in which he "destroyed the posterior part of the spinal marrow by the point of a needle . . . [and] injured the anterior part."[40] The experiments were technically difficult—Bell found it hard to injure one filament without also injuring the other, while getting the spine open without damaging its contents required delicacy and dexterity—but Bell was able to achieve results. He described the injury and destruction of the posterior root as not causing movement, while that of the anterior root caused the animal to shake violently.

It is important that Bell says of these experiments that they were not conclusive, but merely provided encouragement to the belief that his system was correct.[41] Bell's style of physiology required that anatomical and philosophical reasoning precede, and to some extent supersede, vivisection experiments. While he became more explicit about these methodological priorities later in his career, Bell established them from the outset, writing in *Idea of a New Anatomy*, "[i]f I be correct in this view of the subject, then the experiments which have been made upon the brain tend to confirm the conclusions which I should be inclined to draw from strict anatomy."[42] Experiment, then, was used by Bell to support conclusions already deduced from anatomy, much in the same way that he used pathology as a form of natural experimentation. Bell wrote to his brother that "[t]he whole opens up in a new and simple light; the nerves take a simple arrangement; the parts have appropriate nerves; and the whole accords with the phenomena of the pathology."[43] Pathology, usually considered one portion of the Institutes of Medicine (medical theory) and not a part of anatomy or surgery,[44] was adopted by Bell in his physiological reasoning as a way of confirming hypotheses based on anatomy—disfigurement and disease caused anatomical changes corresponding to symptoms that revealed the normal functions of those same parts of the anatomy. His supporters eventually asserted that he was methodologically innovative in this respect, while his detractors claimed he was unscientific. Bell's methodology, tied to his pedagogical program, was as much a part of the priority dispute and competition for scientific credit as were the facts of the discovery itself. It also became a rhetorically useful tool, allowing him to mold himself as opposite to the Frenchman with whom he competed.

Despite his requests for comments on his "little manuscript," Bell received very little attention for the work at the time that he distributed it and was disappointed by the lack of feedback.[45] In the ten years following its printing, Bell continued to work and lecture on the brain and nerves, but did not publish anything on them per se (although he did give a paper before the Royal Society, in July 1821, entitled "On the Nerves," which summarized his work to date and discussed the functions of the fifth and seventh spinal nerves). With the proliferation of cheaply printed texts, publication became an increasingly important measure of good science as the circulation of scientific periodicals expanded dramatically in the 1820s.[46] Thus, when others took up the subject of the nerves a decade after the distribution of Bell's little pamphlet, Bell found his priority disputed and his methods and conclusions under attack.

CONTROVERSY: CHALLENGES FROM ABROAD

Perhaps the strongest threat to Bell's priority came from abroad, from the French physiologist François Magendie. Historians and physiologists have tried to sort out the dispute and to issue credit for the discovery in a variety of ways. L. S. Jacyna and Edwin Clarke pronounced Magendie the victor in terms that are representative of, though perhaps more decisive than, the opinions of many twentieth-century historians and scientists seeking to base priority on "correctness" of facts and methodology,[47] condemning the "incompleteness" of Bell's investigations and his subsequent co-opting of Magendie's later "correct opinions."[48]

Theirs is a particularly striking adjudication because of its clear condemnation of what the authors take to be a violation of modern scientific ethics. Others have been more evenhanded, giving Bell credit for an idea and for an initial theory, while assigning Magendie credit for completing what Bell started.[49] And still others have credited Bell entirely, accusing Magendie of having stolen his idea directly.[50] The predominant view among historians, though, is that Bell only made initial inquiries into the roots of motor and sensory nerves and that Magendie deserves the credit for having proved their functions. What Jacyna and Clarke view as Bell's moral failing—the adjustment and redefinition of the discovery over time—is in fact what is interesting about the priority dispute whose temporality they flatten. In this, as might be expected to be the case in many such long priority disputes, opponents circled round each other and shifted position relative to each other, changing their claims in response both to the other and to the world around them. In this instance, the politics of French-British relations and of medicine and the changing nature of conservative reform of the sort discussed in chapter 3 helped to shift the contours of the dispute itself.

Magendie published his first account of motor and sensory nerves in 1822. In September 1821, John Shaw, who was Bell's nephew and assistant and who knew a little French, had traveled to Paris in order to convey Bell's work to French anatomists. He explained Bell's system to Magendie and, when asked, provided a demonstration on a horse. Shaw had previously only performed the demonstration on an ass and seems to have been confused by what he saw when he cut away the skin on the horse's face: the demonstration did not go as planned and the nerves that Shaw cut failed to cause the expected paralysis of the lip.[51] Still, Magendie found the demonstration intriguing and asked Shaw for a copy of his new laboratory manual and an account of Bell's paper delivered before the Royal Society, both of which he received.

In June 1822, Magendie published an article in the journal he founded, *Journal de physiologie expérimentale et de pathologie,* entitled "Experiments on the Functions of the Roots of the Spinal Nerves."[52] In the article he stated that he had long wanted to try an experiment on spinal nerves but that he had had difficulty opening the spinal cord without killing, or at least seriously injuring, the animal until someone had brought him a litter of eight puppies. The puppies' spinal cords were more malleable, and he had been able to open the vertebral canal without destroying its contents, allowing him to cut first the posterior and then the anterior roots separately and then to sever both together. From these experiments, the first of which produced an animal whose limbs convulsed but were devoid of sensation, the second of which produced flaccid but sensitive limbs, and the third of which produced limbs with neither sensation nor motion, Magendie deduced that "the posterior roots seem to be particularly destined for sensibility, while the anterior roots seem to be especially connected with movement."[53]

After Magendie's article was published, Shaw, who had received a copy of the journal from Magendie himself, wrote to its author to say that Bell had performed the same experiment thirteen years earlier;[54] shortly thereafter, Shaw sent Magendie a copy of Bell's "little book," *Idea of a New Anatomy.*[55] The priority dispute was taking shape.

Magendie's first attempt to settle it appeared immediately in an article that recounted the events to date:

> One sees by this citation of a work which I could not know of, since it had not been made available to the public, that Mr. Bell, led by his ingenious ideas on the nervous system, had been very near to discovering the functions of the spinal roots; nevertheless, the fact that the anterior roots are designed for movement, while the posterior roots belong more particularly to feeling, appears to have escaped him.

Magendie concludes, then, that it was to "having established this fact in a positive manner that I must limit my claims.[56] Magendie seized from the outset on Bell's lack of publication as well as factual errors as the basis of his claims for credit. The changing grounds on which Bell asserted his own priority, meanwhile, give a picture of changes in British medicine generally.

Bell's responses to Magendie were presented in his own later monographs on the nerves; he had not mastered the sort of journal culture that was quickly becoming the norm for scientific publication.[57] The responses came initially in the form of criticisms of Magendie's methodology. He incorporated one of his earliest public reactions to Magendie's work into his *An Exposition of the Natural System of the Nerves* (1825),[58] a volume that detailed Bell's work on the nerves that had been presented before the Royal Society. In it, he wrote:

> In France, where an attempt has been made to deprive me of the originality of these discoveries, experiments without number and without mercy have been made on living animals; not under the direction of anatomical knowledge, or the guidance of just induction, but conducted with cruelty and indifference, in hope to catch at some of the accidental facts of a system which, it is evident, the experimenters did not fully comprehend.[59]

This passage gets to the heart of Bell's critique of Magendie: Magendie's emphasis on "accidental facts," as well as his vicious pursuit of those facts through uninformed vivisection, were deeply flawed. Such claims would have garnered the support of a significant portion of the British medical community at the time, a community that tolerated occasional vivisection but considered it to have limited value.[60] That the first British anticruelty legislation, Martin's Act, was passed in 1822 with the support and testimony of British medical professionals, and that it mentioned Magendie directly, demonstrates the popularity of the antivivisectionist cause in Britain at the time that Bell and Magendie were beginning their dispute.[61] The act, which forbade any person to "wantonly and cruelly beat, abuse, or ill-treat any Horse, Mare, Gelding, Mule, Ass, Ox, Cow, Heifer, Steer, Sheep, or other Cattle,"[62] had little to do with medical research, as it applied only to those who harmed animals of commercial significance belonging to other people (their property) and did not apply to smaller domesticated animals (or to bulls or cocks, who fought for sport), but it did draw attention to vivisection, as arguments made before Parliament in the bill's favor involved testimony about animal experimentation.[63]

In addition to relying on the support of antivivisectionists who objected to the pursuit of such experiments in France, Bell courted additional backing, adopting rhetoric like that of conservative reformers and creating a British tradition and lineage of which he himself was a part. To do so, he emphasized anatomical systems, creating a sense of common purpose and heritage with his contemporaries. Bell often compared his work to the great circulatory system of William Harvey, saying to his brother in a letter in 1819, "[b]elieve me, this is quite an extraordinary business. I think the observations I have been able to make furnish the materials of a grand system which is to revolutionise all we know of this part of anatomy—more than the discovery of the circulation of the blood."[64] It is a telling and ambitious claim, one that makes clear that for Bell, his discovery was more than simply another fact.

By invoking talk of systems in the context of Harvey's discovery, Bell positioned himself within a particular British legacy, alluding to what was recognized as probably the most significant British discovery in anatomy and medicine, attempting to draw together the support of his fellow countrymen for his claims against a foreigner.[65] The French were widely regarded as being empiricists, opposed to "systems," in Britain,[66] and although men like Bell thought their work every bit as empirical as that of the French, they set up an opposition to the French by insisting that facts without systems to explain the relationships of anatomical parts, and without an underlying philosophy, were meaningless. Thus, Harvey became the progenitor of British anatomy and Bell his descendant. Never mind that Harvey vivisected more than Bell did; the nationalistic argument was something of a bricolage. The consistency in the philosophies of the antimaterialist, anti-French, British anatomists resided in the importance ascribed to unity and underlying purpose, or function, in anatomical structures—to systems.[67]

Bell's cause, like that of the antivivisectionists, took on nationalistic tones, such that Bell even wrote, in an 1823 letter to his brother, "[y]ou may send for the 'Medical Journal'—the last number of the yellow book—if you please, where you will find some strictures in my favour and against the French. They, you know, have accused me of taking from them!"[68] He spoke of the French and of Magendie almost interchangeably during the early years of the dispute, partly because Bell trusted that his fellow Britons would share his opinions about the French and their style of medical science. It is, perhaps, not surprising that a man who practiced during the Napoleonic Wars would assume that fellow British doctors would rally around him if his territory was being threatened by the French. But because the flow of British medical students studying in Paris, which had slowed to a trickle during the Napoleonic Wars, increased significantly once the wars ended,[69] Bell found the medical community around him—its methods and its sympathies—shifting.

Thus Bell continued to insist that Magendie had stolen his idea and that Magendie had then pursued it in a way that was methodologically inappropriate, but his attacks became more specific and more detailed, no longer taking for granted that other British anatomists were opposed to vivisection or to the French. Just a few years after the unequivocal passage above about experiments "without number and without mercy," Bell claimed, in an 1828 lecture before the College of Surgeons, that his experiments "upon the fifth nerve, and the seventh, were repeated before him [Magendie]; that the rationale of these experiments was explained to him; that he had a little work put into his hands, in which these experiments upon the roots of the spinal nerves were described. . . ." Even so, Bell charged, "I am constrained, in this place, to say that he may not have understood these experiments upon the seventh, or on the fifth . . . that he may even, in short, have employed his fingers, those 'pickers and stealers,' as Shakespeare calls them, without the control of his head—without intention or ideas of any kind—with a perfect purity that belongs to entire ignorance."[70] Bell's vitriol was undiminished, but this time he equated Magendie's experiments with his own. He did so because he could no longer assume that his audience disavowed the methods of the French; that is, by this point he seems to have assumed that they did not want to hear him rail against vivisection on moral grounds.

Just a year later, in a letter to the editor of the *London Medical Gazette*, Bell seems to confirm that experimental physiology was taking hold in Britain, writing, "How often shall I have to make an apology for not believing in the opinions of experimenters?"[71] He went on to argue that he tried precisely the same experiments that Magendie did "with every assistance possible. . . . My experiments on this subject entirely failed. . . . When, therefore, twelve years afterwards, I addressed the Royal Society, I put all these experiments aside, and founded my reasoning upon that which was not only correct but was easily ascertained to be so."[72] Magendie's work, Bell said, was subject to mistakes—the procedures Magendie followed could not possibly have allowed him to distinguish between sensory and motor nerves.[73] Here, Bell refined and combined his earlier arguments—Magendie stole his experiments, but those experiments don't work, so Magendie must also have stolen his results. With a subtle shift in rhetoric, Bell made himself the original thinker but also the expert on a methodology that he had earlier condemned, one that had become increasingly popular among his colleagues, and he did so, fittingly, in the sort of vehicle that had become an indispensible part of scientific publishing in the late 1820s, the weekly journal.

Bell's argument changed several more times within his lifetime. In the next revision of the argument, in 1834, he praised Johannes Müller for deciding the controversy experimentally, saying: "[h]e has repeated the experiments with the utmost care, insulating the distinct roots, and observing the effects when they are variously irritated. He has shewn that by experimenting upon frogs, the conclusions which I had announced are confirmed in a manner which admits of no question or doubt; and that one root—the anterior—is for motion alone, and the posterior for sensation alone."[74] Bell's acceptance of experiments as conclusive in a debate in which he had rejected experimentation as improper and unsuccessful from the outset could be seen as merely convenient—Bell favored experimenters when they helped his cause. Surely there was some of that self-interest at work here, but Bell was also responding to a general shift in the way that the British medical community viewed experiments on living animals (although it is not clear whether he himself changed his view of such practices). In 1834, one British surgeon testified before a parliamentary committee on medical education that "[t]he proper course of physiology is that taught by experiment,"[75] while at the same time, British scientists were launching a "science in decline" argument,[76] claiming that all the good scientific work was being done in Germany and France where the state provided resources and where there was not the same public scrutiny of vivisection.[77] In order to maintain the support of his colleagues, Bell needed to incorporate some of the new epistemology, one in which animal experimentation and the pursuit of what Bell earlier called "accidental facts" provided the foundation of physiological knowledge, into a defense of his own priority. Bell's changing appeals to colleagues reflect the chronology of the debate, which had come to stand in as a debate between modes of medical science—priority disputes that go on for years probably often reflect changes in the communities that make up the audience for those disputes.

The final twist in the argument between Bell and Magendie came in the form of an admission of error on Bell's part. In his 1834 clinical lecture on diseases of the nerves of the head, published in the *London Medical Gazette*, Bell stated:

> My experiments proved the portio dura to be the nerve of motion to all the muscles of the side of the face, with the exception of the muscles of the jaws; . . . with regard to the lips, I was led into a mistake in my first experiments, which Magendie corrected. I thought that the lips, besides obtaining the power of motion principally from the branches of the portio dura, were also, to a certain degree, under the control of branches prolonged from the motor root of the fifth pair: and this I conceived was for the purpose of associating the lips

> and the cheeks in the combined actions of mastication. I was in error as to the particular branch which is so prolonged to the cheeks—an error into which I should not have fallen, had I examined with more care, before my first experiments, the anatomy of the roots of the fifth pair, as it is given in several of the best German authors . . .[78]

Here, again, Bell was posing as an experimentalist and allying himself with German anatomy. But this was a unique admission of a mistake by Bell that did not minimize Magendie's work. Still, it ends with Bell's insistence on the importance of anatomy *before* experimentation, demonstrating that although Bell might have recognized a place for experimentation within physiology, he was still committed to the primacy of what he called "higher anatomy."[79] He thought this superior to a strictly experimental physiology, even though programs like his were going out of fashion with his contemporaries.

Bell's changing defense of his priority in the dispute with Magendie suggests the changes that Bell perceived in the British medical community—his audience and potential allies in the dispute. As the first chapters of this book describe, by the outset of the debate in 1822, Bell's career and ambitions had been forged in and achieved through teaching, and the discovery had become a part of his classroom-centered research and publicizing of his work. Potential patrons—social, professional, and natural philosophical—had been given a copy of a little book, printed at the cost of the anatomist in a small print run. That was how one established greatness at the time. Publication in the grandest of forms—an engraved folio—would be the capstone of his career, published toward the end of his lifetime for all posterity.[80] At the time the dispute broke out, broader British medical and surgical communities were located across a multitude of hospitals and small schools like Bell's, but those communities were still local and personal and could be united against a common enemy by disputes like Bell's that resonated with broader political concerns, much as they were united around conservative reform, as a means of preventing revolution. Thus, in the initial years, Bell responded aggressively and directly to Magendie, drawing on nationalistic and anti-French sentiment and employing the rhetoric of conservative reform and British traditionalism, hoping to unite other British medical men in his cause.

But by the latter half of Bell's teaching career, it became clear that simple nationalism would not win allies for Bell and that he would have to change his approach in order to bring together a necessary broad swath of supporters, and more importantly, in order to impress the newly cosmopolitan British medical scientists. Although wider medical communities were brought together by larger and broader medical schools and new medical journals that circulated widely, the community itself became, perhaps, more clearly divided into camps defined not by national or nationalistic politics, but by local and professional divisions, educational backgrounds, and varied positions on reforms that were being instituted. Audiences were no longer local in the same way, and medical science was no longer primarily constructed and conveyed in the classroom. With such a fragmenting audience, Bell was fighting a losing battle. Bell and his students carried forward the dispute with Magendie with vigor, but the dispute only really piqued the attention of a British medical community that was divided along generational and political lines when Bell's former student, Herbert Mayo, claimed the discovery for himself and also allied himself methodologically with Magendie.

CONTROVERSY: CHALLENGES FROM WITHIN GREAT BRITAIN

Bell's response to Magendie's assertion of priority can be contrasted directly with his response to his own countryman and student, Herbert Mayo. As discussed above in chapter 3, during the early part of the nineteenth century the British medical community was becoming publicly factionalized. These factions developed partly through the growth of periodicals with political slants, [81] but they also grew through discussions over licensing and educational reform.[82] Bell's priority dispute with British contemporaries can be understood as both evidence for and a product of these professional politics.

While a few others of Bell's British contemporaries took different sides in his debate with Magendie, in Bell's view the most brutal betrayal by one of his own countrymen came from his own student, Herbert Mayo. Mayo studied with Bell from 1812 to 1815, first at the Great Windmill Street School of Anatomy and later in the wards of the Middlesex Hospital.[83] Bell's and Mayo's careers remained closely intertwined, as was often the case with ambitious pupils and their teachers in early nineteenth-century London. Mayo became a house surgeon alongside Bell at the Middlesex Hospital in 1818, and in 1826 he and another of Bell's students, Caesar Hawkins, bought the Great Windmill Street School of Anatomy from Bell. When the Middlesex Hospital School was founded in 1835 as a rival to London University, Charles Bell and Herbert Mayo were both on the surgical faculty, and Dr. Francis Hawkins, brother of Bell's student Caesar Hawkins, was on the faculty of physicians.[84] As is clear, both from such a web of connected careers and from letters like those Bell himself wrote about cultivating professional support in his early days in London, teachers and patrons found their students, relatives, and friends positions and helped them to gain a foothold in a competitive medical marketplace.

Like Bell, Mayo worked on the nerves, and he clearly took many of Bell's approaches. Both worked on the ass, and experiments they performed were similar. Furthermore, both were known for their skillful drawings, and in fact they are notable as the only two authors whose lectures, printed in the opening issues of the *London Medical Gazette*, were printed with illustrations, signaling the importance of such illustrations to their classroom teaching.[85] But Mayo differed from Bell in his assignment of functions to the fifth and seventh cranial nerves and disputed Bell's theory of respiratory nerves. Their disagreement remained civil at first—in the first edition of Mayo's *Outlines of Human Physiology*, published in 1827, Mayo credited Bell with developing the experiments from which he and François Magendie separately worked, saying, "But when thus sharing the claim to these discoveries between M. Magendie and myself, I should in justice state that the experiments in each case were but improvements on those which Mr. Bell had previously performed."[86] But the dispute quickly turned vicious, as is evident in Mayo's omission of the above passage crediting Bell in later editions of the book. The conflict surely escalated at least in part because Mayo not only claimed correct assignment of the facial nerves for himself but also declared that Magendie's claims to priority of discovery of the roots of motor and sensory nerves were valid.[87]

The dispute between student and teacher is also revealing of changes in the professional nature of British medicine. Herbert Mayo was a part of an ambitious group of medical practitioners—his father, Dr. John Mayo, was a physician who became a governor at the Middlesex Hospital, and his older brother, Thomas, was a physician as well, having graduated from Oxford. Thomas Mayo inherited his father's practice, was elected a fellow of Oriel College, and served as president of the College of Physicians, leaving Herbert with much to live up to.[88] Herbert Mayo's family provided him with a thorough education (he studied at Leiden, where he took an MD, as well as at the Middlesex Hospital and with Bell at the Windmill Street School) and also with many valuable connections. But Herbert Mayo was regarded by his contemporaries as being particularly eager for advancement, perhaps in a vulgar sort of way. When King's College Medical School was opened in 1831, for example, Mayo was elected to the Chair of Anatomy and Physiology on the basis of his reputation as a well-educated, skilled practitioner. According to an 1852 report of Mayo's death and eloge in the *Provincial Medical and Surgical Journal,* however, when, in 1836, "the same chair was vacated at University College by the resignation of Dr. Jones Quain, Mr. Mayo proffered himself as one of the candidates but was unsuccessful with the additional annoyance of having excited feelings of distrust in the breasts of the Council of King's College."[89] This sort of move, accepting one teaching post and then trying secretly to get a better one when it opened up, was thought to be overly aggressive and ungentlemanly. The editors of the *London Medical Gazette* advised Mayo at the time, "We shall be plain with Mr. Mayo: This over-vaulting ambition of his is both pitiful and ridiculous: in our opinion, instead of scampering about, as he seems so strongly disposed to do, he should confine himself to his chair of surgery, and be thankful that he has got it."[90] His obituary in the *Lancet* in 1852 even declared him "somewhat conceited."[91] It was exactly the sort of uncouth image that Bell, always aware of the need for patronage, avoided at all costs. But Mayo's career was different from Bell's; he attained a university professorship at an early stage in what was becoming a genuine career with a predictable path. With his foreign education, his connection to physicians (rather than surgeons), and his family's social standing, Mayo's conflict with Bell took on the added dimension of professional politics, encompassing the tensions between a group of older and perhaps more conservative surgeon-anatomists educated in Britain who sought both the standing and the patronage of gentlemen and a group of young medical scientist-surgeons who had been educated abroad and were baldly and perhaps rudely ambitious.

BELL, MAYO, AND THE NERVES OF THE FACE

In Bell's dispute with Magendie, facial nerves gave Bell's method of physiological inquiry particular trouble. His limited experiments did not always yield the expected results, as was the case when John Shaw demonstrated Bell's work on the facial nerves in front of Magendie. The facial nerves became the center of Bell's dispute with Mayo as well. At around the same time that Magendie began publishing on the nerves (1822), Herbert Mayo, who had been Bell's demonstrator at the Great Windmill Street School and thus was intimately familiar with Bell's work, published his *Anatomical and Physiological Commentaries,* in which he directly disputed much of Bell's work on the facial nerves. The facial nerves were complex and intertwined, with many ganglia, unlike the spinal nerves; consequently, dissecting the facial nerves was not a straightforward process. They also provided much of the pathology (including what we now know as Bell's Palsy) that Bell used to deduce the function of anatomical structures: if he cut one nerve to relieve pain or tension in a particular portion of the face, or if someone had an injury that resulted in the paralysis of particular facial muscles, Bell could use that to confirm his physiological theories. Mayo, who did a good deal of work on the facial nerves, found their complicated structures good for making the case for the necessity of vivisection.

In 1821, Bell published an article entitled "On the Nerves; Giving an Account of Some Experiments on Their Structure and Functions, Which Lead to a New Arrangement of the System" in the *Philosophical Transactions of the Royal Society of London*, in which he divided nerves into two categories: a symmetrical system of nerves and superadded or irregular nerves. He focused particularly on the trigeminus, or fifth pair of cranial nerves, and the facial, or seventh pair of cranial nerves (figures 15 and 16), declaring that the fifth pair belonged to the symmetrical system, while the portio dura of the seventh pair (which was divided at the time into the portio dura and the portio mollis[92]) belonged to the superadded, or respiratory, nerves. The fifth pair was important to Bell, essentially because it resembled the spinal nerves—Bell called the fifth pair "the spinal nerves of the head"—and conversely, the scheme of spinal nerves was important because it explained the complicated fifth pair of cranial nerves. According to Bell's scheme, the fifth pair had sensory branches that, like the ganglion-filled sensory nerves of the spinal cord, emerged from a ganglion, while also having a small motor root that bypassed the ganglion. He traced the origin of this discovery to the process of teaching and demonstrating in front of his students.[93] The idea of a symmetrical system of nerves and a superadded system of nerves helped Bell to explain both the similar paths of two nerves that appeared to have motor functions (which would be a redundancy not plausible to someone who favored the elegance of a designful Creator) and the pathology of partial paralysis of the face—nerves from the symmetrical system would produce paralysis of voluntary motion whereas those of the superadded system would paralyze respiratory (and therefore involuntary) functions.[94] He often told his audiences that such paralysis was as good as an experiment, demonstrating on a living patient how chewing motions could remain undisturbed, due to an uninjured fifth nerve, while at the same time the patient could not laugh or control his facial expression. Thus, according to Bell, the portio dura was the respiratory nerve of the face and "all those motions of the nostril, lips, or face generally, which accord with

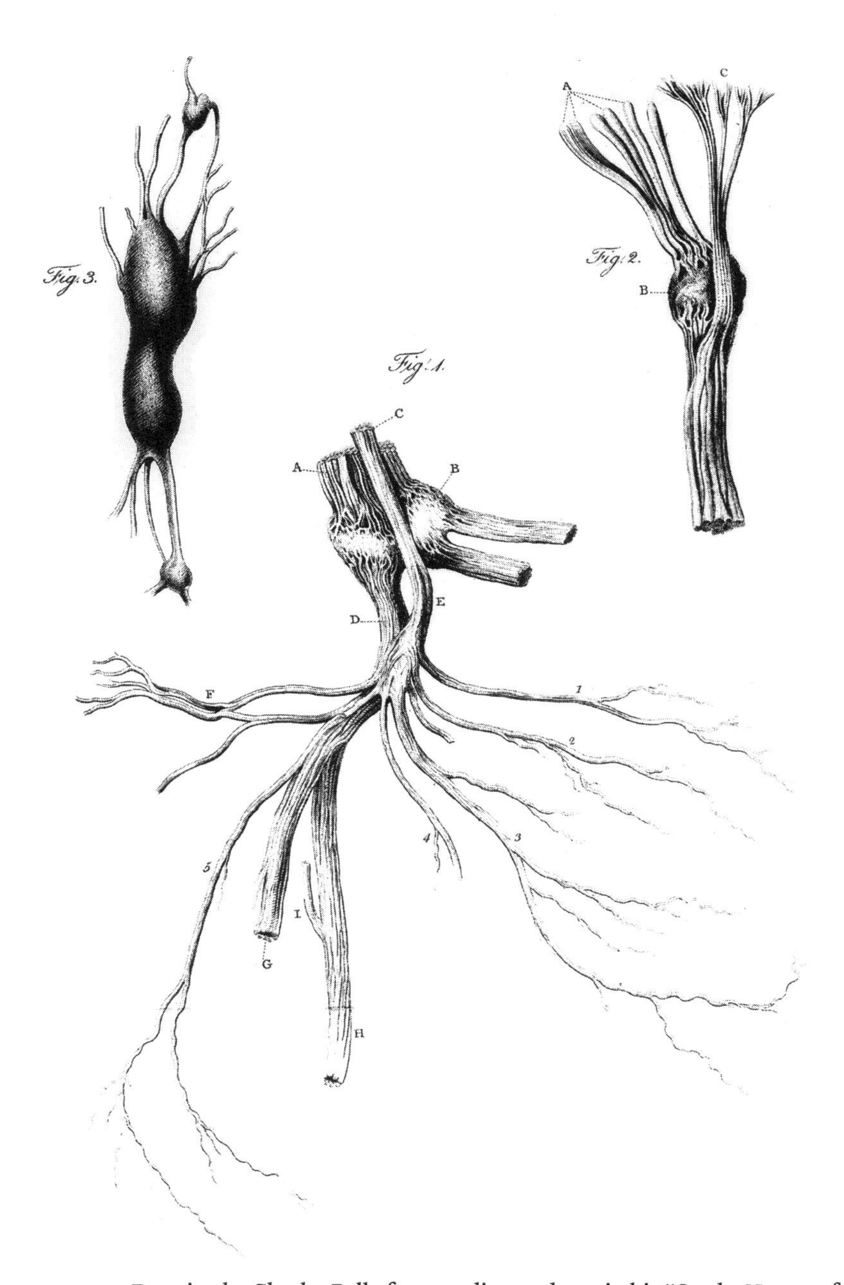

FIGURE 15. Drawing by Charles Bell of nerves dissected out, in his "On the Nerves of the Face," *Philosophical Transactions of the Royal Society of London* 119 (1829), plate viii. The large nerve in the middle of the page is the fifth nerve, dissected out. It shows both the ganglion and the motor root passing in front of the ganglion. The smaller nerve above and to the right of the fifth nerve is the ganglion of a spinal nerve, placed in the drawing to show its similarity to the ganglion of the fifth nerve. The nerve to the left of the fifth nerve is the ganglion of a sympathetic nerve and is meant to show the difference between ganglia of the sympathetic nervous system and those of the symmetrical system (of which the fifth nerve and the spinal nerves were a part). Image courtesy of the Wellcome Library, London.

FIGURE 16. Engraving of nerves of the face in Charles Bell, "On the Nerves of the Face," *Philosophical Transactions of the Royal Society of London* 119 (1829), plate ix. In this image, drawn by Bell, *A*, which is right under the ear, is the portio dura at its origin, with its principal branches cut. *B*, which is pinned down in front of the face, is the trunk of the portio dura, the seventh nerve, while *C*, pinned immediately next to it, is the third branch of the fifth nerve, which joins the portio dura, showing the complexity of the interwoven nerves of the face.

the motions of the chest in respiration, depend solely on this nerve," as did the muscles of expression, which Bell believed to be related.[95] Without a functioning portio dura, the parts of the face could not coordinate with the lungs or produce expression (which Bell considered to be mostly involuntary).

Herbert Mayo rejected Bell's system of respiratory nerves, as well as his assignment of functions to the fifth and seventh cranial nerves. When cutting the fifth cranial nerve, Mayo noted no loss of muscle tone, but on cutting the facial or seventh cranial nerve "the lips immediately fell away from the teeth, and hung flaccid, and the nostrils lost all movement."[96] Thus, where Bell had declared the fifth nerve a motor nerve because when he cut it the ass appeared unable to eat, Mayo concluded that it was a sensory nerve and that the animal did not eat with its lips because it could not feel the food; when the food was placed on its tongue, it could still devour its oats. Similarly, Mayo found that Bell's experiments on the seventh nerve did not go far enough: when cutting *both* sides of the seventh pair of cranial nerves, Mayo found that not only respiratory, but all motor functions ceased, leading Mayo to declare the seventh pair a general motor nerve. Caesar Hawkins, Bell's student and coproprietor with Mayo of the Great Windmill Street School, summed things up in his 1849 Hunterian Oration to the College of Surgeons: "Here, too, Sir Charles Bell's humanity stood in his way for he only divided the portio dura on one side of the face, the division of which by Mayo on both sides left no doubt that no power of motion was derived from the fifth to the muscles of the face, and it was soon acknowledged by everyone that the seventh was their sole motor nerve."[97] Here, as with Magendie, vivisection as a method of inquiry was at the heart of the dispute.

By 1829, the debate had gotten fierce enough that Mayo and Alexander Shaw, Charles Bell's nephew, engaged in a heated exchange in the *London Medical Gazette*.[98] Mayo wrote disparagingly of Shaw, saying that his letter to the *Gazette* "has produced a very painful impression upon my mind. It is painful to witness the adoption, at the very outset of life, of a course so misguided."[99] Mayo then went on to describe Bell's theory of the respiratory nerves and quoted Magendie concurring with Bell, implying that Magendie also believed in a system of "superadded" nerves. This allowed Mayo to set himself up as the true discoverer of the functions of the fifth and seventh nerves. He declared that this notion of "respiratory nerves"[100] had always given him difficulty but that he felt it unnecessary to discuss the system further, as the theory was "falling into disrepute" and he wanted the reputation not of being the one to debunk another discovery but of being the one to determine the proper functions of the portio dura and of the facial branches.[101]

Charles Bell did not answer these accusations himself, but Alexander Shaw did so at great length, presumably in part on Bell's behalf. He began by attacking Mayo's "air of condescension":

> I believe I am not altogether destitute of that ancient virtue which enforces respect and deference to the aged; but let me ask what was the example afforded me by Mr. Mayo himself when he was a young man, commencing his professional career—eight years ago? What was his conduct towards Mr. Bell, his senior by many years—his teacher, and in whose house he had resided? It is entertaining to hear Mr. Mayo, of all men, assuming the tone he does—he who commenced by opposing in the most reckless manner all that Mr. Bell had done on the subject of the nerves, and who afterwards claimed as his own the most essential and prominent parts of his preceptor's discoveries.

"He was protected from Mr. Bell's animadversions," Shaw summarized acidly, "merely because he was a very young man, and had been his house pupil."[102] Mayo had failed to show the deference expected of a student toward his teacher; he had bad professional manners. In addition, Shaw asserted that Mayo had omitted discussion in his *Commentaries* of both Bell's work and John Shaw's work on the nerves and that, if he had had true evidence regarding their falseness, he would have demonstrated that he had repeated previous experiments and found them lacking. Finally, Shaw demanded that Mayo, who had said that if priority were to be allocated to Bell or to Magendie, he preferred Magendie, give some reason for his choice.[103]

Although Shaw wrote extensively in response to Mayo, Bell clearly felt that it was inappropriate for him to respond to Mayo directly. Shaw acted as Bell's representative in this controversy much as Samuel Clarke did for Isaac Newton and T. H. Huxley would do later for Darwin.[104] But by the end of the exchange Bell felt it necessary to distance himself from the whole mess, allowing the editors of the *London Medical Gazette* to write: "We are authorized, by Mr. Bell, to contradict the insinuation that he is the concealed opponent in the controversy between Mr. Mayo . . . and Mr. Shaw. He has neither written nor dictated any thing on the subject in dispute."[105]

Bell never did directly address Mayo's betrayal either in published work or in surviving private letters. His career and Mayo's followed very similar trajectories, and they often worked at the same institution, making civility a necessity. But Bell's initial reluctance to attack Mayo directly might also have had to do with Bell's hope of uniting British anatomists and medical practitioners behind his own anatomical physiology and against French physiology. Bell wrote in his 1830 edition of *The Nervous System of the Human Body* (which was written at approximately the same time as the dispute with Mayo) about his difficulty in keeping his "pupils to the examples of our own great countrymen."[106] The British traditions created and invoked by Bell and others of the conservative reform movement, along with the idea that British medical science and pedagogy were best served by following in the traditions of their great predecessors and disavowing continental experimental physiology, were being displaced by a new cosmopolitanism. The *London Medical Gazette* reviewer of Bell's book concurred about the new enthusiasm for foreign science, saying, "The nationality which displays itself in this just appeal, cannot, we repeat, be too much admired and encouraged. It is, in truth, full time for all rational thinkers to be heartily tired of that rage which is so prevalent in favour of foreign opinions."[107] But whereas Bell was very outspoken about Magendie's claims to discovery, Bell refrained from attacking his own countryman and student by name. Bell, who was always building alliances and trying to promote his anatomical physiology as a particularly British endeavor, would not have benefited by attacking another British physiologist, especially not one whom he had trained himself.

Mayo, on the other hand, continued to write on the nerves and to dispute Bell's findings. In 1834 he wrote in a letter to the editor of the *Medical Quarterly Review* that "Magendie, by ingeniously using very young animals in his experiments, succeeded in obtaining a positive result, and in realizing the discovery, which is honestly his," describing the discovery of the difference in function between spinal nerves with and without ganglia. And in case it was not enough to credit Magendie, Mayo discredited Bell, saying, "To Sir Charles Bell's various publications, in which he claims or assumes credit for discoveries to which he is not entitled, the following words of Seneca would form an excellent motto: 'ISTA PRO INGENIO FINGUNTUR, NON EX SCIENTIAE VI.'"[108]

In 1839, after Bell had returned to Edinburgh and had grown increasingly distant from the London medical scene, Alexander Shaw published *Narrative of the Discoveries of Sir Charles Bell in the Nervous System*,[109] an extensive defense of Bell's priority. In it, Shaw focused almost entirely on attacking Mayo, reserving only one of six chapters for Magendie. Much of his case against Mayo had to do with establishing the dates on which various players were working on aspects of the discovery, the extent to which Mayo's work was done under the supervision or at the instigation of Charles Bell, and

Mayo's personal conduct toward his mentor, but he also approached the dispute from a methodological standpoint, discussing vivisection at length and redefining, yet again, the substance of Charles Bell's innovation.

Shaw's explanation of why vivisection was a method inappropriate to physiological research on the nerves demonstrated both similarities to Bell's views on the matter and clear differences. It is apparent that Shaw was arguing against a majority of physiologists in Britain, most of whom were vivisectionists or at least accepted vivisection by this time (which was not the case during the height of Bell's career), and Shaw argued the point, therefore, on technical and not ethical grounds. He described in rather excruciating detail the process whereby experiments on the nerves would be conducted, with the physiologist cutting through skin, muscles, and bones, and introducing bone scissors into the vertebral canal "to tear and break up the fragments, and disclose parts contained within." He asked rhetorically: "can it be supposed that, after suffering from the tortures of such a proceeding, there is any animal, however submissive to the infliction of pain or high in its courage that could endure the further and concluding parts of the experiment with such a degree of patience as to admit of correct observations being made in regard to the amount of sensibility appertaining to either of the roots?" Could we judge whether, in the case of an animal whose spine marrow has been cut up, "whether its struggles and cries result from the severity of the wound inflicted, or depend on the fresh injuries that we commit on the roots of its nerves?"[110]

Shaw claimed that there was no way that the nerve roots could be left exposed and the animal unaltered, and that, although one could demonstrate which roots were responsible for motion through vivisection, it would be impossible to show which were responsible for sensation. But he did not go so far as to suggest that vivisection was always unacceptable; it simply was not effective in studying the nerves. Shaw explained that it was for this reason that Bell was unable to assign the property of sensation to either root experimentally, even though he did mention that the anterior root alone was capable of exciting the muscles to contract. In order to determine which was the root of sensation, Bell had to return to anatomy and to the argument, based in natural theology, that the human body is an elegant and purposeful system without unnecessary redundancies. According to this argument, circulation of nervous impulses, much like the circulation of the blood, would make logical sense, and such a system would require that the posterior root be for sensation, so that one root carried the will of the brain out to the body while the parallel root returned sensory perceptions to the brain.

After declaring vivisection inappropriate for investigations of the nerves, thereby undermining the methods of Bell's detractors, Shaw went on to make the case that Bell's discovery was fundamentally methodological: that Bell's innovation lay in the sheer act of focusing on the roots of the nerves, and that this was far more significant than whatever Magendie or Mayo did afterward.

> Here, then, is the simple explanation of the principle on which all these new discoveries have been based. It consists, I repeat, in supposing that, to investigate the functions of the nervous system successfully, we must devote our attention, not to the trunks, as was formerly done, but to the roots of the nerves. Accordingly, whoever was the first to suggest and follow out that new method of prosecuting the subject, must be declared the true originator of the recent improvements in this department of physiology. It is by the test of who did the most to establish this law, that we must decide to whom we are indebted for these discoveries.[111]

By shifting from a defense of Bell on the basis of his results to a defense based on approach, Shaw redefined the terms of the debate, denying Mayo and Magendie priority by definition and making their work appear derivative. In a sense, this defense allowed for the support of experimentalists and vivisectionists, for as Shaw pointed out, without the focus on the roots, there could be no revealing experiments on the nerves. Shaw's account also placed vivisection early in the narrative of Bell's work—Bell discovered the function of one root by vivisection and then resorted to philosophical principles to assign a function to the other root—rather than presenting it as a simple means of confirming a scheme already worked out. Shaw's defense of Bell was written for an audience of medical scientists different from that for which Bell's own work had been written. Where Bell assumed his early audience was made up largely of anti-French and antivivisectionist surgeon-anatomists, Shaw was writing to convince physiologists educated in the style of continental medical scientists. We can even see the use of the word "physiology" as a sort of shorthand for some of these generational differences. For Bell, physiology was just a sort of anatomy with function and movement added, but for Mayo, Alexander Shaw, and the next generation of medical men, physiology was one of the new experimental medical sciences.

Bell regarded anatomy as the true basis of physiology and used the word "physiology" with caution. In his 1830 letter written to his students upon his resignation from the London University, Bell wrote, "To those who know how little I value physiology, in the common acceptation of the term, it will be a proof of my desire to see the experiment of the new school fairly tried, that I submitted to be called professor of a science (if a science it be) on which an inceptor candidate for medical degrees would read lectures more readily than I could."[112] He had a clear distaste for a discipline that was associated with vivisection and the French and a new style of medical scientist who failed to develop a proper grounding in anatomy (figure 17). Bell continued his letter to his students by saying, "You are aware that the subjects on which I lectured were the higher departments of anatomy-that I reasoned on a demonstration in which my knowledge of anatomy and my experience of disease came into use as laying the just principles in the practice of your profession."[113] Higher anatomy—akin to natural philosophy, built on dissection and on reason, elegant in its outcomes, amounting to more than simple (sometimes contradictory) facts amassed together, developed and propagated in the classroom—was the foundational *British* medical science, according to Bell.

Mayo, however, following Magendie, was of a new generation of medical practitioners and scientists, locating himself within a new discipline of physiology, which he considered to be a science in its own right, with its own experimental methods that would help him to understand the living body.[114] Physiology was a lab science and a journal science. It would not be performed in the classroom (as dissection) and in the home of its writer (as philosophizing in higher anatomy). It would not be published in grand, illustrated folios. This new generation called themselves "physiologists" to distinguish themselves from older anatomists like Bell. Herbert Mayo called his two most famous works *Anatomical and Physiological Commentaries* and *Outlines of Human Physiology*, just as François Magendie called the journal that he founded the *Journal de physiologie expérimentale et pathologie*. Even Alexander Shaw, Bell's strongest advocate, was a part of the new generation of practitioners and used the term physiology without reservation, even applying it to Bell's work in the quotation cited above: "Accordingly, whoever was the first to suggest and follow out that new method of prosecuting the subject, must be declared the true originator of the recent improvements in this department of physiology."[115] Bell's work on the nerves and the priority dispute that followed took place against the backdrop of a medical community in transition and a medical science in the making. The changes in Bell's claims to discovery help to demonstrate the changes in the audience to which he was appealing.

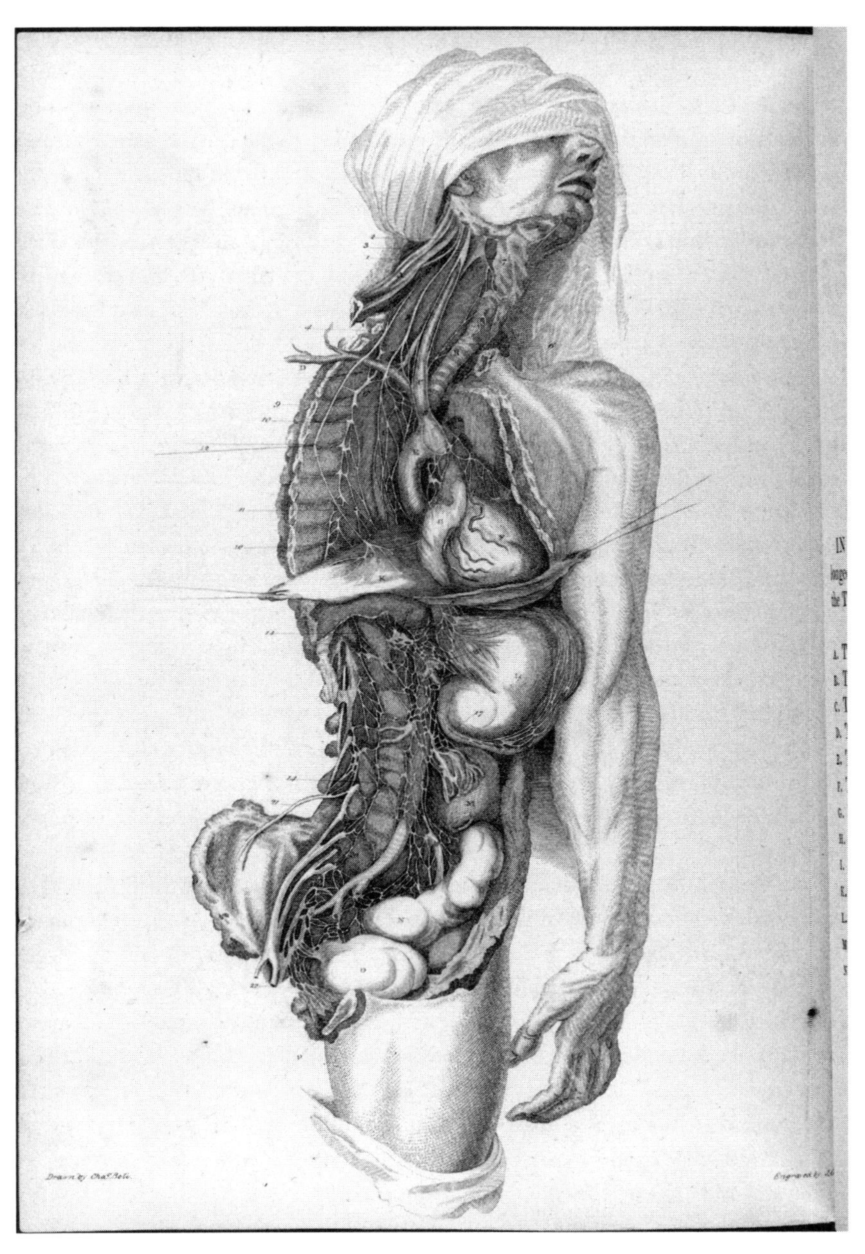

FIGURE 17. Engraving of a dissection of the human body, based on a drawing by Charles Bell for his *Anatomy of the Brain* (London: Longman, Hurst, Rees, and Orme, 1802). Image courtesy of the Wellcome Library, London.

FROM ANATOMISTS TO PHYSIOLOGISTS

When Bell first moved to London, surgeon-anatomists made their way by cultivating patrons among a local community, ran small schools of anatomy in their homes, and taught students in hospitals that provided clinical experience for practically-minded students. Natural theology was in vogue. The Napoleonic Wars limited exchanges between British and French medical practitioners,[116] and it was safe to assume that British practitioners would reject the vivisection that was being adopted in France as a method for understanding the body. In this environment, Bell began to work on the nerves, imagining both a specialist and nonspecialist audience for "his discovery," which he would present to the scientific community through lectures at his Windmill Street School of Anatomy, thereby drawing a larger number of students who would help generate income.

By the time Bell left London to return to Edinburgh in 1836, London had universities, offering career paths to enterprising medical men; journals abounded, making the London scene far less local; and physiology had supplanted anatomy as the fundamental medical science. Physiology was not tied to (or tied down by) clinical experience or application in the way that anatomy and pathology were, and it did not rely on cadavers or on shared philosophical assumptions about a creator or design. What had seemed ethically problematic at the beginning of the nineteenth century was now accepted as a routine part of medical education (even if its acceptability to nonmedical society in Britain was just starting to be seriously questioned).[117]

In order for Bell to defend his discovery, he had to reposition his claims to accord with the trends in British medical science. Bell's rhetorical use of antivivisection waned over the first third of the nineteenth century along with his emphasis on natural theology as the fundamental basis of his discovery. When Bell arrived in London, he was a part of a cohort of great surgeons and medical doctors, and of great teachers, that had made their way to the capital of practical medical education to teach in the numerous charitable hospitals and small schools of anatomy that London offered. This group, disproportionately Scottish, regularly dined together, worked together, and traveled in the same non-medical intellectual circles. By the time Bell left the city, the small schools had been driven out in part because large universities and hospitals had supplanted them.[118] The medical community had now clearly splintered into "scientists" and "practitioners," both of which were separate from other emerging disciplines in the sciences

(Bell's idea about spreading his discovery by lecturing on it to Banks's coterie of women would have seemed entirely untenable by the 1840s).[119] The most ambitious medical scientists would have conducted and taught physiology in the style of the French or Germans, as an experimental laboratory science and not as a fundamentally practical or clinic-based science.

In this environment, the politics that had determined earlier professional alliances—nationalism, birthplace, and the identity of one's teachers and relatives—were to some extent replaced by alliances built on place in the medical reform movement and other sorts of professional politics. Journals, which were overtly aligned with professional causes and with men of particular medical sects and backgrounds, demonstrated professional rank and politics visibly. In such a landscape, Bell attempted to fashion a nationalistic science of anatomy integrated with relevant pathology; that is, pathology that revealed the normal functions of anatomical structures,[120] to form what he called "higher anatomy." His higher anatomy was a science that used experiment but did not rely on it, and he propagated that science in the classroom and through the close, small social circles and networks of patronage that ruled in the early part of the nineteenth century. While his discovery of a system of the nerves was recognized both within Britain and abroad, its underpinnings—a philosophical anatomy based in natural theology, demonstrative anatomical preservations and drawings that were revealing of the systems of the living body, and a science based in pedagogy—were not similarly recognized. Bell's struggle to assert his priority occurred alongside a parallel and intertwined struggle to establish the sort of medical and surgical education upon which Bell built his own reputation, a struggle that ultimately failed. That world of private medical schools was quickly changing, in part through the advent of medical journals and the audiences they helped to create.

Epilogue

Ultimately, it probably did not matter terribly much that Bell did not *win* his priority dispute. The dispute itself brought him fame, and over a decade after it began in earnest, he was knighted, so it is hard to make the claim that Bell suffered personally for lack of a decisive victory—though he himself was quite resentful about its progression. In 1830, in the introduction to a collection of his assembled writings on the nerves for the Royal Society, Bell wrote bitterly:

> The early announcement of my occupations failed to draw one encouraging sentence from medical men . . . To myself this has ceased to be of consequence; but I confess, I regret to leave those young men who have honourably and zealously assisted me in these inquiries, in the delusive hope of laboring to the gratification of their own profession—the pleasure arising from the pursuit of natural knowledge, and the society of men of science, must be their sufficient.[1]

His was the unhappiness of a man who had been left behind, and while perhaps Bell was not defeated in his dispute with Magendie, somehow he felt himself to have lost.

What was actually lost during the period of the dispute was a unified, classroom-based medical science of anatomy. Historians have periodically examined the places and spaces of science in an attempt to understand its practices or its invisible technicians or its material culture.[2] The story of Charles Bell's priority dispute and the rise of modern medical science could be told as one about the spaces of science as well. In early nineteenth-century London, anatomy was a science of the classroom. That was its space. Looking back at the dispute between Bell and Herbert Mayo, in particular, shows how that space was broken apart by the end of the period in question.

Lacking regular career paths, natural-philosophically minded medical men of the late eighteenth and early nineteenth centuries established schools. In those schools, they performed research. That research was fostered through teaching; house pupils learned from their masters but assisted with research as well, which was conducted not only through dissections but also through the preservation and drawing of specimens. Classroom drawings and specimens housed in the school's museum constituted proofs of new discoveries.

The classroom was also the space in which research was publicized. New ideas became marketing tools to bring students (and even colleagues) in the door, their fees paid to the lecturer, thereby helping to sustain a scientific enterprise lacking state or other institutional funding. Charles Bell lectured on his new discovery for over a decade while he refined his work before publishing it in the *Philosophical Transactions*, and eventually in a folio edition of six papers delivered before the Royal Society with an added introduction, titled *The Nervous System of the Human Body: Embracing the Papers Delivered to the Royal Society on the Subject of the Nerves* (1830). That the new medical journals of the 1820s could sustain themselves by filling their weekly publications primarily with the contents of medical lectures suggests that this practice of teaching one's new research was widespread. By the time of Bell's death in 1842, that space had been divided.

The classroom had been the center of "conservative reform" in London, much as it had been the center of all else in Bell's world. Conservative reformers had argued for the strength of London's practical training, which was focused on therapeutics. That training took place in classrooms where the best teachers were rewarded by a competitive system and in charitable hospitals where the primacy of the healing (as opposed to scientific) mission of medical men was made clear by the boards that governed them. Conservative reformers saw opportunities to improve on this style of education, while keeping its basic character, by systematizing coursework, improving credentialing examinations, and reframing licenses and degrees so that they better reflected the realities of practice (as well as the varieties of courses offered in London and, indeed, the coursework that many students undertook). But whatever change was to come, conservative reformers would have it that the ideals of such change were "British" in nature. Such gradual and practically oriented conservative reform lost out, as did the notion that particularly British ideals should pertain to any reforms enacted. By the time of Bell's death, London University had been peacefully reformed along radical, continental lines, and with it went the rest of London. The fate of two separate reform movements can be traced through Bell's priority dispute and its aftermath, showing that, although Bell did not exactly lose, his reform movement did.

Sir Charles Bell died in a town called Hallow in Scotland after another of his attacks of abdominal pain. He was sixty-eight years old. In the end, his brother George only survived him by about a year. After his death, Marion Bell, Charles's widow, wrote of those things that were most important to her husband, devoting part of that remembrance to his teaching: "He disliked the display of oratorical power. He desired to be easily intelligible,—and still giving a dramatic and picturesque grace. He was often overheard practising his voice in particular passages . . . He varied his lectures year by year, and studied for each as if it were to be the only one."[3] Bell's pupils, who assumed great significance in both his personal and professional lives, were an important part of his legacy. He was remembered for the things he cultivated—a reputation as a teacher being foremost among them. Bell, who was prone to reflection, had in 1835 written a passage that seemed destined to capture his legacy:

> I have seen enough to satisfy me of what the world can offer a man—I mean this great world; and were you to look back to my letters, you would find the opinion uniformly expressed that the place of a professor who fills his place is the most respectable in life. My hands are better for operation than any I have seen at work; but an operating surgeon's life has no equivalent reward in this world . . . I must be the teacher and consulting surgeon to be happy.[4]

But academic standing, built for Bell in the classroom, had become the province of the laboratory scientist by the time Bell's career had ended.

That central space, the classroom, which had encompassed all of Bell's scientific activities—reputation-building, research, teaching, and publicizing—and that had served as a base for his local network of colleagues and patrons, had been shattered into several pieces by the time Magendie and Mayo were contesting Bell's discovery. By the 1820s, research and teaching were conducted as separate endeavors. These changes can be seen in Mayo's claims to priority. Mayo's work, which evidently began as a part of Bell's instruction of house pupils, in his school and under his direction, was in Bell's view a product of the workshop and its leader. But for Mayo, the experimental work was his own, since he had done it with his own hands. The correct performance of the experiment was its crucial element.

As with Magendie, the science itself was in the determination of the correct fact.

In his *Anatomical and Physiological Commentaries* (1822), Herbert Mayo wrote that the only way of knowing the functions of the nerves of the face had been to take advantage of a few chance instances of injuries to those nerves. Experiment on animals, and extrapolation from those experiments to the functions of nerves in humans, he then pronounced, was the only way forward. After detailing an experiment on an ass, Mayo reflected, "I was induced to perform the preceding experiments on reading an essay by Mr. Bell, in which a novel view of the functions of certain nerves is propounded; resting in part upon experiments in great measure similar to those above narrated, but differing materially in their results."[5] Mayo therefore conceded the similarity of his experiments to Bell's, but to do so for Mayo was not to express indebtedness. After all, Bell's experiments proved worth very little when he got the wrong results: "Sir C. Bell was carried by these experiments very near to the truth, but he failed at that time to ascertain it: he inferred from his experiments, indeed, that the anterior and posterior roots of the spinal nerves have different functions; but in the nature of those functions he was mistaken."[6] One might note the impersonal and detached tone of Mayo's discussion of his teacher's work, and, in fact, of his account of how he came to know about Bell's experiments. The text nowhere suggests that Mayo was Bell's pupil and that their relationship in the classroom was a formative one. He went on to write about his own contribution to studies on the nerves in ways that are suggestive of what Mayo thought constituted discovery, in a passage that is entirely about experiments and facts (experimental proof).[7]

One can see the devaluing of the classroom in its very absence in these accounts, and in the disregard of Mayo for his teacher, but the classroom's loss of status becomes even more apparent when one considers Alexander Shaw's defense of his uncle and mentor. In his 1839 *Narrative of the Discoveries of Charles Bell*, Shaw wrote: "Mr. Mayo, before entering on these inquiries, had enjoyed the advantage of being Sir Charles Bell's pupil, and had been admitted, both by this gentleman and Mr. Shaw, to pursue these researches along with them."[8] Shaw makes the intertwined relationship of teaching and research clear—as Bell's pupil, Mayo would have been involved in research *with* Bell and Shaw. According to Shaw, he also "had been employed by Sir Charles Bell to make preparations for the museum of Great Windmill Street, with the view of illustrating the difference in the structure of the roots of the nerves."[9]

Mayo's science did not have a place for an all-encompassing classroom.

His work was not made up of ideas and systems, refined and passed on in a workshop-like school. His science was a set of facts built in a laboratory, and Bell's facts were simply wrong. Of vivisection, physiology's experimental method, Mayo wrote, "Harvey deduced from experiment proofs of the circulation of the blood, and almost every important addition to physiology has been obtained by the same method."[10] The idea for an experiment was virtually irrelevant. It was its execution that mattered. Mayo was a part of a new cohort, the heirs to radical reform. He was continentally trained (in Leiden, after he had finished at Bell's school), and he admired continentals. His work was based on science of the sort that became established at London University, the sort opposed by conservative reformers.

Work done within a community but not published was similarly unworthy of scientific credit in this new world of medical science, and publication's contours had changed. Mayo wrote of his textbook, *Outlines of Human Physiology*, that it was made up of the "heads" from his lectures, essentially offering only a basic outline of the systems of the human body and their physiology, but that "[a] treatise fitting the magnitude of the subject would not suit the present thriving condition of Medical Science." The progress of physiology had become so fast-paced "that the first part of such a work would begin to be obsolete before the publication of the last; not to mention, that it would be too voluminous and expensive for students, and would be neglected by men advanced in the medical profession, for the scattered original essays from which it must have been compiled."[11] Mayo's medical science was a rapidly changing one in which novelty was contained in essays (journal articles) that were published long before books could be. Again, the dispute makes clear the contours of reform. The *Lancet* and other medical journals were no longer the place for the contents of lecture and advice to young students; now they were regarded as the place for original research. Those periodicals, which had first sought to constitute a new community by creating specialized social groups within London's medical men, dividing them along political lines, found their specialized social group instead among the "researchers" who published on medicine and its experiments.

Mayo discusses the place of the classroom in such a medical science, saying, "Physiology can only be adequately taught in Anatomical Theatres, when the Lectures of each year contain the discoveries of the preceding . . ."[12] The classroom was no longer where discovery was built, nor even where original discovery was conveyed—it was for last year's now-certain science. Pedagogy had lost its pride of place.

This book has traced conservative reform in medicine through the life of Charles Bell, a man who cobbled together networks of social and professional patronage during an era when anatomy lacked a "career path." Pedagogy permeates this story about a reform movement whose politics were individual, flexible, and patchwork, rather than ideological and univocal. The classroom was at the center of British conservative reform, and in fact of all of medical science during this period, providing a unifying space in which research, teaching, and publicizing happened all together. By the mid- to late nineteenth century, none of this would pertain. Without the threat of revolution, there was no need for a separate branch of reform, "conservative reform."

Gerald Geison's classic treatment of Michael Foster's research school at Cambridge[13] demonstrates just how different Britain had become within thirty years of Bell's death, just a generation or two after his own heyday. Geison quotes Foster as writing, "If you have enough pocket money, don't spend a moment longer on teaching than you can help . . . a few years hence you will only have the ideas of your youth to fall back upon. Do *now* as much original work as you possibly can and let everything else go as it likes."[14] And he goes on to talk about how Foster seemed to follow the very trend that J. B. Morrell described when he wrote, "if a [research] school wanted a more than local reputation it had to publish its work. The relatively easy access to publication opportunities, or best of all control of them, enabled a school to convert private work into public knowledge and fame."[15] This world was far removed from that of Charles Bell when he wrote that "the place of a professor who fills his place is the most respectable in life," or when he told his brother he would not publish his discovery, but instead would make the classrooms of London "ring with it." The subject of anatomy had changed as well. It lost physiology as a handmaiden and derivative science. Again, Foster demonstrates the preeminence of a physiology standing on its own in the founding of his *Journal of Physiology*, meant in title and subject to "formally dissolve the traditional union between anatomy and physiology in England."[16]

Sir Charles Bell's name is tied to a priority dispute, a law regarding the roots of motor and sensory nerves, and a medical disorder (Bell's palsy). He was knighted and made a professor at the school at which he started, the University of Edinburgh. His career came full circle—after finding himself without options in Edinburgh because of his brother's cantankerous nature and disputes with highly placed medical men, Charles went to London, where he developed a professional and social network, a flexible set of politics and alliances to suit the company he was in, and the appearance of a gentleman. In her remembrances after his death, Charles's wife Marion quoted him as saying often, "You see, I am not a quarrelsome fellow! There is not a man in my profession with whom I am not good friends!"[17] Those friends and patrons helped to fill his classrooms, within which he taught a discovery of the sort he himself yearned to find and set out to uncover. With such a discovery in hand, he could make up for his educational deficits—his lack of training in natural history[18]—and could consider himself a colleague of Sir Joseph Banks or Sir Humphry Davy, a philosophical gentleman like them. And in the end, having made his career from a hodgepodge of accomplishments and efforts, having taught surgery and anatomy, having established schools and published papers, Bell could assume a position that a more stable sort of career path might have allowed him at the outset. Such a career path was available to Michael Foster, just one example of the professional physiologists who were trained in the decades after Bell. It was to one of Bell's own schools, University College, London, that Michael Foster went for his medical training, and there he became a lecturer and then a professor, before going on to a praelectorship in physiology at Cambridge and eventually a newly created chair of physiology in the university. Career paths altered the social and political makeup of London medicine, as well as its emphasis on teaching.

Charles Bell's priority dispute might in fact have helped to secure his fame. But he was among the last of a generation of Enlightenment-era medical men who formed their careers, their research, and their publications through the private classrooms of early nineteenth-century London; whose ambitions for reform were fundamentally about conserving something quintessentially British; and whose politics were shaped by the exigencies of developing a living through various kinds of patronage at a time when careers in medical science simply did not exist. Within a decade or two that world was gone. Professionalization and regularized education, the ambitions of reformers, had been realized, along with regular career paths, and with that change, the classroom had shattered, its functions divided among other spaces, each with its own audience and function: the laboratory, the clinic, and the classroom. They are the spaces of modern medicine, the ones we recognize today, and we see them as the hallmark of medical science. But for a brief window of time, conservative reformers living in an Age of Reform tried to fashion a particularly British science of medicine, stressing systems of education rooted in philosophical anatomy and traditions of humane science rooted in practice, in London's competitive, private classrooms.

Acknowledgments

Over ten years, one accumulates many debts. First, I am grateful for the institutional support of the Chemical Heritage Foundation, my employer over the past four years, and the place where most of this book was written. I also owe thanks to those who helped to sponsor or foster my research while I was a graduate student: the faculty and staff of the Cornell Science & Technology Studies department, the National Science Foundation (for Dissertation Improvement Grant 0646371), the Mario Einaudi Center (for a Luigi Einaudi Fellowship for Research in Europe), and the Philadelphia Area Center for the History of Science, where I found not only a Dissertation Research Fellowship and an office near to my newly transplanted home but also a new academic community and a mentor and friend in Babak Ashrafi.

Family, friends, and colleagues over the years have played a role, whether by reading or listening to parts of the book or simply by providing reassurance. For that support, I'd like to thank my parents (particularly my mother, who made sure that the manuscript was as readable as I had hoped); members of my dissertation committee, Stephen Hilgartner, Suman Seth, and Rachel Weil; and also Eva Åhrén, Nico Bertoloni Meli, Tom Broman, John Christie, Michael Dennis, Michelle DiMeo, Matthew Eddy, Mary Fissell, Robert Fox, Jan Golinski, Rob Kohler, Tayra Lanuza Navarro, Mike Lynch, Anna Maerker, Christine Nawa, Lynn Nyhart, Lissa Roberts, Mike Sappol, Emily Stanback, John Harley Warner, Audra Wolfe, and Nasser Zakariya.

I'm also grateful to Jen Lentz, Charlotte Markey, Lynne Schofield, Gina Stack and their children for helping me to find happy escapes from academic work and for perspective on what's most important in life.

Karen Darling at the University of Chicago Press deserves thanks for believing in this project, for helping to walk me through the submission process when I needed it, and for her patience when work and life put me behind schedule.

I owe much to my colleagues at the Chemical Heritage Foundation. Ron Brashear and Rebecca Ortenberg gave me the flexibility and assistance that made it possible to finish writing a book while also running the Beckman Center, many of my coworkers provided friendship, and the fellows who spent time in the Beckman Center over the last four years (2010–14) often gave counsel or lent moral support.

I want to express special gratitude to David Caruso, without whom this book project would never have been possible.

And finally, it remains only to acknowledge Peter Dear and Theo Caruso, in their rather opposite roles. Peter provided inspiration and the sort of encouragement that sustained me, even when it seemed most improbable that I would succeed. He read every word of this book, questioned some of them and praised others, and it is a better book because he did. Theo has no idea what is in this book: he knows I am a doctor of the sort who doesn't help people, but that I write about people who were the sorts of doctors who did help people; he knows that sometimes we all got more "screen time" in our household because I needed to be on my laptop to finish something; and most importantly, he knows that on the day I mailed my pile of paper off to Chicago, he got to eat dinner AND dessert at a restaurant. He's the best reason I know to try to succeed at anything, and I hope one day this makes him proud.

Notes

INTRODUCTION

1. Charles Bell, *Letters of Sir Charles Bell, Selected from his Correspondence with his Brother George Joseph Bell* (London: J. Murray, 1870), 324 (October 12, 1831).

2. Ibid., 325 (October 12, 1831).

3. Sir John Herschel's letters and diaries have been published in several different editions. His life was celebrated in a bicentennial symposium held at a school bearing his name, and he has been the subject of quite a few biographies, including Günther Buttmann and David S. Evans, *The Shadow of the Telescope: A Biography of John Herschel* (Guildford: Lutterworth Press, 1974); Steven Ruskin, *John Herschel's Cape Voyage: Private Science, Public Imagination, and the Ambitions of Empire*, Science, Technology, and Culture, 1700–1945 (Burlington: Ashgate, 2004); John F. W. Herschel and David S. Evans, *Herschel at the Cape: Diaries and Correspondence of Sir John Herschel, 1834–1838*, History of Science Series 1 (Austin: University of Texas Press, 1969). His classmate at Cambridge, Charles Babbage, has received even more attention. For a small sample, see Maboth Moseley, *Irascible Genius: The Life of Charles Babbage* (Chicago: H. Regnery, 1970); Doron Swade and Charles Babbage, *The Difference Engine: Charles Babbage and the Quest to Build the First Computer*, 1st American ed. (New York: Viking, 2001); Anthony Hyman, *Charles Babbage, Pioneer of the Computer* (Oxford: Oxford University Press, 1984); and Bruce Collier and James H. MacLachlan, *Charles Babbage and the Engines of Perfection*, Oxford Portraits in Science (New York: Oxford University Press, 1998).

4. Astley Cooper, *The Lectures of Sir Astley Cooper, Bart. on the Principles and Practice of Surgery*, 3 vols. (Boston: Wells and Lilly, 1825), 184–93.

5. Matthew Kaufman, *Robert Liston: Surgery's Hero* (Edinburgh: Royal College of Surgeons of Edinburgh, 2009).

6. Cooper, *Lectures of Sir Astley Cooper*, 185.

7. Reginald Magee, "Surgery in the Pre-Anaesthetic Era: The Life and Work of Robert Liston," *Australian and New Zealand Society of the History of Medicine* 2, no. 1 (2000): 125–26.

8. Bell, *Letters*, 10.

9. John Struthers, *Historical Sketch of the Edinburgh Anatomy School* (Edinburgh: Maclachlan and Stewart, 1867), 37–44; and Christopher Lawrence, "The Edinburgh Medical School and the end of the 'Old Thing,' 1790–1830," *History of Universities* 7 (1988).

10. Struthers, *Historical Sketch of the Edinburgh Anatomy School*, 40–41.

11. John Bell, *Letters on Professional Character and Manners: On the Education of a Surgeon, and the Duties and Qualifications of a Physician: Addressed to James Gregory, M.D.* (Edinburgh: John Moir, 1810).

12. "John Bell," in *Penny Cyclopedia of the Society for the Diffusion of Useful Knowledge* (London: Charles Knight, 1835), 4:190.

13. Ibid.

14. Bell, *Letters*, 338 (1834).

15. Ibid., 64, footnote.

16. Reviewer, "Letters and Discoveries of Sir Charles Bell," *Edinburgh Review* 136, no. 1872 (1872): 207. It's interesting that Sydney Smith includes John Millar, the Glasgow professor of law, philosophical historian, political theorist, pupil of Adam Smith, and Whig-Reformer, along with Edinburgh luminaries. It suggests the strength of a Scottish philosophical community that transcended geography.

17. Bell, *Letters*, 128 (August 5, 1808).

18. For just a few examples of historical work on the significance of the brain or mind as a subject of study during the period, see Alan Richardson, *British Romanticism and the Science of the Mind*, Cambridge Studies in Romanticism (Cambridge: Cambridge University Press, 2001); Paul F. Cranefield and Charles Bell, *The Way In and the Way Out: François Magendie, Charles Bell, and the Roots of the Spinal Nerves: With a Facsimile of Charles Bell's Annotated Copy of His Ideas of a New Anatomy of the Brain* (Mount Kisco: Futura Publishing, 1974); Roger Cooter, *Phrenology in the British Isles: An Annotated Historical Biobibliography and Index* (Metuchen: Scarecrow Press, 1989); John Van Wyhe, *Phrenology and the Origins of Victorian Scientific Naturalism* (Aldershot: Ashgate, 2004); and Steven Shapin, "The Politics of Observation: Cerebral Anatomy and Social Interests in the Edinburgh Phrenology Disputes," in *On the Margins of Science: the Social Construction of Rejected Knowledge*, ed. Roy Wallis, Sociological Review Monographs (Keele: University of Keele, 1979).

19. Thinking was by no means uniform, of course, among these men. But close-knit circles of philosophically and natural philosophically minded men seem to have produced common themes of inquiry across disciplines and generations. For a very good overview of the Scottish Enlightenment, see Alexander Broadie, ed., *The Cambridge Companion to the Scottish Enlightenment* (Cambridge: Cambridge University Press, 2003), esp. chap. 2 ("Religion and Rational Theology," by M. A. Stewart), chap. 3 ("The Human Mind and its Powers," by Alexander Broadie), and chap. 5 ("Science in the Scottish Enlightenment," by Paul Wood).

20. Christopher Lawrence, "Medicine as Culture: Edinburgh and the Scottish Enlightenment" (PhD thesis, University of London, 1984), chap. 8.

21. John Locke, *An Essay Concerning Human Understanding* (London: printed by Eliz. Holt for Thomas Bassett, 1690), book 2, chap. 9. William Cheselden, also a London surgeon, addressed it, e.g., in W. Cheselden, "An Account of some Observations made by a young Gentleman, who was born blind, or lost his Sight so early, that he had no Remembrance of ever having seen, and was couch'd between 13 and 14 Years of Age," *Philosophical Transactions* 402 (1728): 447–50. Cheselden applied what might be seen as the first experimental evidence to the problem when he treated a congenitally blind boy whose sight was restored when his cataracts were removed. The boy did not know shapes and could not distinguish things by sight.

22. Dugald Stewart, *Elements of the Philosophy of the Human Mind*, vol. 3 in *The Collected Works of Dugald Stewart* (Edinburgh: Thomas Constable and Co., [1827] 1854), 212. This sort of thinking was also found among French Enlightenment philosophers like Étienne Bonnot de Condillac.

23. Quoted in Amédée Pichot, *The Life and Labours of Sir Charles Bell* (London: R. Bentley, 1860), 172.

24. The Royal College of Surgeons had been known for its corruption and financial mishaps since the eighteenth century. See Bernice Hamilton, "The Medical Professions in the 18th Century," *Economic History Review* 4, no. 2 (1951).

25. For accounts of British medicine that describe a shift toward scientific medicine in the late nineteenth century, see, e.g., Andrew Cunningham and Perry Williams, *The Laboratory Revolution in Medicine* (Cambridge: Cambridge University Press, 1992); and Terrie M. Romano, *Making Medicine Scientific: John Burdon Sanderson and the Culture of Victorian Science* (Baltimore: Johns Hopkins University Press, 2002).

26. Roy Porter, "Medical Lecturing in Georgian London," *British Journal for the History of Science* 28, no. 1 (1995): 99.

27. Susan C. Lawrence, "Entrepreneurs and Private Enterprise: The Development of Medical Lecturing in London, 1775–1820," *Bulletin of the History of Medicine* 62 (1988); idem, "Private Enterprise and Public Interests: Medical Education and the Apothecaries' Act, 1780–1825," in *British Medicine in an Age of Reform*, ed. Roger French and Andrew Wear (London: Routledge, 1991); and idem, *Charitable Knowledge: Hospital Pupils and Practitioners in Eighteenth-Century London* (Cambridge: Cambridge University Press, 1996).

28. For more on the Great Reform Act, see, e.g., Arthur Burns and Joanna Innes, *Rethinking the Age of Reform: Britain 1780–1850* (Cambridge: Cambridge University Press, 2003); and Eric J. Evans, *The Great Reform Act of 1832*, Lancaster Pamphlets (London: Methuen, 1983); idem, *Britain Before the Reform Act: Politics and Society 1815–1832*, Seminar Studies in History (London: Longman, 1989).

29. Burns and Innes, *Rethinking the Age of Reform.*

30. Adrian J. Desmond, *The Politics of Evolution: Morphology, Medicine, and Reform in Radical London* (Chicago: University of Chicago Press, 1989).

31. Jan Golinski, "Humphry Davy: The Experimental Self," *Eighteenth-Century Studies* 45, no. 1 (2010).

32. Michael D. Gordin, *A Well-Ordered Thing: Dmitrii Mendeleev and the Shadow of the Periodic Table* (New York: Basic Books, 2004), 6.

33. For more on French medicine in the period, see, e.g., John E. Lesch, *Science and Medicine in France: The Emergence of Experimental Physiology, 1790–1855* (Cambridge: Harvard University Press, 1984); Toby Gelfand, *Professionalizing Modern Medicine: Paris Surgeons and Medical Science and Institutions in the 18th Century* (Westport: Greenwood Press, 1980); J. M. D. Olmsted, *François Magendie, Pioneer in Experimental Physiology and Scientific Medicine in Nineteenth Century France* (New York: Schuman's, 1944); Russell Charles Maulitz, *Morbid Appearances: the Anatomy of Pathology in the Early Nineteenth Century*, Cambridge History of Medicine (Cambridge; New York: Cambridge University Press, 1987); John Harley Warner, *Against the Spirit of System: The French Impulse in Nineteenth-Century American Medicine* (Princeton: Princeton University Press, 1998); and L. W. B. Brockliss and Colin Jones, *The Medical World of Early Modern France* (Oxford: Oxford University Press, 1997).

CHAPTER ONE

1. Bell, *Letters*, 148 (May 1809).

2. Ibid., 20–21 (November 30, 1804).

3. Here Bell is much like his contemporary, Humphry Davy, who was relentlessly self-fashioning and ambitious. See Golinski, "Humphry Davy: The Experimental Self," 17.

4. Desmond, *Politics of Evolution*, 93.

5. Charles wrote to George: "His speech was not good in effect, tho' in manner and expression admirably well adapted to the hearers. He had the same slips of words, and returning to correct his expressions, as you must recollect; energetic and plausible, from an argumentative manner. But there was on no occasion loud or long applause . . . The people were very indifferent, and took it as a good joke to hear Fox speak." Bell, *Letters*, 45 (May 4, 1805).

6. Ibid., 172 (March 12, 1810).

7. Ibid.

8. Ibid.

9. Nicholas Jewson, "Medical Knowledge and the Patronage System in 18th Century England," *Sociology* 8, no. 3 (1974).

10. For more on the private schools and hospitals of London during this period, see Lawrence, "Entrepreneurs and Private Enterprise," "Private Enterprise and Public Interests," and *Charitable Knowledge*.

11. Bell, *Letters*, 27 (December 22, 1804).

12. Ibid., 36 (January 31, 1805).

13. Ibid., 53 (July 5, 1805).

14. Carin Berkowitz, "Medical Science as Pedagogy in Early Nineteenth-Century Britain: Charles Bell and the Politics of London Medical Reform" (PhD dissertation, Cornell University, 2010), 204.

15. Bell, *Letters*, 20 (November 30, 1804).

16. Maulitz, *Morbid Appearances*.

17. H. B. Donkin and C. MacNamara, *The Westminster Hospital Reports*, vol. 1 (London: J. E. Adlard, Bartholomew Close, 1885), 11.

18. Stewart Craig Thomson, "The Surgeon-Anatomists of Great Windmill Street School," *Bulletin of the Society of Medical History of Chicago* 5 (1937–46).

19. John Leonard Thornton, *John Abernethy: A Biography* (London: printed for the author; distributed by Simpkin Marshall, 1953).

20. R. C. Brock, *The Life and Work of Astley Cooper* (Edinburgh: E. & S. Livingstone Ltd., 1952).

21. Bell, *Letters*, e.g., 23, 28, 50, 53, 54, 71, 74, 95,101, 49, 50, 52, etc.

22. Michael Bevan, "Cline, Henry (1750–1827)," *Oxford Dictionary of National Biography*, Oxford University Press (2004), http://www.oxforddnb.com/view/article/5673.

23. Bell, *Letters*, 74 (June 27, 1806).

24. Ibid., 38 (March 7, 1805).

25. Ibid., 46 (May 18, 1805).

26. Ibid., 40 (March 23, 1805).

27. Ibid., 46–47 (May 18, 1805).

28. Ibid., 119–21 (November 1807) and 29–31 (January 1808).

29. Aileen Fyfe, *Science and Salvation: Evangelical Popular Science Publishing in Victorian Britain* (Chicago: University of Chicago Press, 2004), 43; and Alan Rauch, *Useful Knowledge: The Victorians, Morality, and the March of Intellect* (Durham: Duke University Press, 2001), 41.

30. Bell, *Letters*, 54 (September 24, 1805).

31. Ibid., 43 (April 19, 1805).

32. Ibid., 23 (December 4, 1804).

33. Ibid., 276 (February 1, 1822).

34. Ibid., 112 (October 1807).

35. Ibid., 45–46 (May 4, 1805). Or, in another instance, "Everything goes well with me but money. I confess to you, my dear George, I am sick and heavy, and out of heart, at being so poor" (ibid., 217 [May 16, 1814]).

36. Lawrence, *Charitable Knowledge*.

37. Bell, *Letters*, 24 (December 7, 1804).

38. Ibid., 69.

39. Ibid., 70.

40. Ibid., 65 (February 4, 1806).

41. Ibid.

42. Ibid.

43. Ibid., 100 (July 6, 1807).

44. For more on nineteenth-century medical museums and their audiences, see Samuel J.

M. M. Alberti, *Morbid Curiosities: Medical Museums in Nineteenth-century Britain* (Oxford: Oxford University Press, 2011); and Samuel J. M. M. Alberti and Elizabeth Hallam, *Medical Museums: Past, Present, Future* (London: Royal College of Surgeons of England, 2013).

45. Bell, *Letters*, 55 (September 30, 1805).

46. Ibid., 63 (January 20, 1806).

47. Lawrence, "Entrepreneurs and Private Enterprise," 182.

48. Ibid., 181.

49. Gordon Gordon-Taylor and E. W. Walls, *Sir Charles Bell, His Life and Times* (Edinburgh: E. & S. Livingstone, 1958), illustration facing 41.

50. Bell, *Letters*, 63 (January 20, 1806).

51. Ibid.

52. Ibid., 65 (February 8, 1806).

53. Ibid., 66 (February 8, 1806).

54. Ibid.

55. Ibid., 101 (August 17, 1807).

56. Ibid., 111 (October 1, 1807).

57. Ibid., 112 (October 1, 1807).

58. Ibid., 113 (October 14, 1807).

59. Ibid., 71 (April 7, 1806).

60. Ibid., 117 (December 5, 1807).

61. Charles Bell, *Essays on the Anatomy of Expression in Painting* (London: Longman, Hurst, Rees, and Orme, 1806); *A System of Operative Surgery: Founded on the Basis of Anatomy*, 2 vols. (London: Longman, Hurst, Rees, and Orme, 1807); *Idea of a New Anatomy of the Brain* (London, 1811); and *Letters Concerning the Diseases of the Urethra* (London: Longman, 1810).

62. Bell, *Letters*, 117–18 (December 5, 1807).

63. Notably Herbert Mayo, Alexander Shaw, and Caesar Hawkins. Thomson, "Surgeon-Anatomists of Great Windmill Street School."

64. Bell, *Letters*, 156 (October 2, 1809).

65. Ibid., 183 (February 1811).

66. Ibid., 184 (February 1811).

67. Ibid.

68. Ibid., 69 (February 11, 1806).

69. Ibid., 51 (July 1, 1805).

70. Ibid.

71. Bell, *Essays on the Anatomy and Philosophy of Expression* (London: John Murray, 1824), 6. Note that Bell had his 1806 *Essays on the Anatomy of Expression in Painting* reprinted under this slightly different title in 1824. Page references are to the 1824 edition unless otherwise noted.

72. Bell, *Letters*, 52 (July 5, 1805).

73. Ibid.

74. Jack Morrell and Arnold Thackray, *Gentlemen of Science: Early Years of the British Association for the Advancement of Science* (Oxford: Oxford University Press, 1981).

75. Bell, *Letters*, 53 (July 5, 1805).

76. Ibid., 96 (May 21, 1807).

77. Ibid., 91 (March 26, 1807).

78. Bell, *Essays on the Anatomy and Philosophy of Expression*, vi.

79. Bell, *Letters*, 119 (December 5, 1807).

80. For more on Joseph Banks, see John Gascoigne, *Joseph Banks and the English Enlightenment: Useful Knowledge and Polite Culture* (Cambridge: Cambridge University Press, 1994); and idem, *Science in the Service of Empire: Joseph Banks, the British State and the Uses of Science in the Age of Revolution* (Cambridge: Cambridge University Press, 1998).

81. Bell, *Letters*, 119 (December 1807).

82. Ibid., 131 (October 21, 1808).

83. Bell sent a copy to Fuseli, who apparently sent a card in return thanking Bell for his "rich and truly valuable" gift. Bell, *Letters*, 80 (August 25, 1806).

84. Ibid., 122 (January 1, 1808).

85. Ibid., 120 (November 26, 1807).

86. W. F. Bynum, "Carlisle, Sir Anthony (1768–1840)," *Oxford Dictionary of National Biography*, Oxford University Press (2004), http://www.oxforddnb.com/view/article/4687.

87. Bell, *Letters*, 121 (January 1, 1808).

88. Frederick Cummings, "B. R. Haydon and His School," *Journal of the Warburg and Courtauld Institutes* 26, no. 3/4 (1963); and Aris Sarafianos, "B. R. Haydon and Racial Science: the Politics of the Human Figure and the Art Profession in the Early Nineteenth Century," *Visual Culture in Britain* 7, no. 1 (2006).

89. He wrote in 1808, "By the bye, my patients are at this time very few, but I am easy on that score." Bell, *Letters*, 134 (November 26, 1808).

90. Bell, *A System of Operative Surgery*, iv.

91. Ibid.

92. Ibid.

93. Bell, *Letters*, 26 (December 15, 1804).

94. Ibid., 39 (March 7, 1805).

95. Ibid., 72 (April 14, 1806).

96. Ibid., 119 (December 5, 1807).

97. Ibid., 149 (May 27, 1809).

98. Ibid., 142 (February 1809).

99. Ibid., 139 (February 3, 1809).

100. Ibid., 140 (February 1809).

101. Ibid., 147 (May 23, 1809).

102. Ibid., 176 (June 9, 1810).

103. Ibid., 147 (May 23, 1809).

104. Ibid., 73 (May 19, 1806).

105. Ibid., 97 (May 21, 1807).

106. Francis Jeffrey was an Edinburgh lawyer and member of the Speculative Society. He was also friends with Henry Brougham and, with Sydney Smith and Francis Horner, was among the founding members of the *Edinburgh Review* in 1810. He was later its editor. For more, see Michael Fry, "Jeffrey, Francis, Lord Jeffrey (1773–1850)," in *Oxford Dictionary of National Biography*, Oxford University Press (2004), http://www.oxforddnb.com/view/article/14698.

107. Bell, *Letters*, 96 (May 21, 1807).

108. Steven Shapin, "Phrenological Knowledge and the Social Structure of Early Nineteenth-Century Edinburgh," *Annals of Science* 32, no. 3 (1975), and "The Politics of Observation: Cerebral Anatomy and Social Interests in the Edinburgh Phrenology Disputes," in Wallis, *On the Margins of Science: The Social Construction of Rejected Knowledge*, Sociological Review Monographs 27 (Keele: University of Keele, 1979); Lisa Rosner, *Medical Education in the Age of Improvement: Edinburgh Students and Apprentices* 1760–1826 (Edinburgh: Edinburgh University Press, 1991); Richardson, *British Romanticism and Science of the Mind*; Van Wyhe, *Phrenology and Origins of Victorian Scientific Naturalism*; and Roger Cooter, *The Cultural Meaning of Popular Science: Phrenology and the Organization of Consent in Nineteenth-Century Britain*, Cambridge History of Medicine (Cambridge: Cambridge University Press, 1984).

109. Bell, *Letters*, 118 (December 5, 1807).

110. Ibid., 124 (July 8, 1808).

111. Warner, *Against the Spirit of System*.

112. Michael Barfoot, "Philosophy and Method in Cullen's Medical Teaching," in *William Cullen and the Eighteenth-Century Medical World*, ed. A. Doig, J. P. S. Ferguson, I. A. Milne, and

R. Passmore (Edinburgh: Edinburgh University Press, 1993); and J. R. R. Christie, "The Origins and Development of the Scottish Scientific Community, 1680–1760," *History of Science* 12 (1974).

113. Lawrence, "Medicine as Culture."

114. J. R. R. Christie, "Ether and the Science of Chemistry: 1740–1790," in *Conceptions of Ether: Studies in the History of Ether Theories*, 1740–1900, ed. Jonathan Hodge and Geoffrey Cantor (Cambridge: Cambridge University Press, 1981); and Arthur Donovan, *Philosophical Chemistry in the Scottish Enlightenment* (Edinburgh: Edinburgh University Press, 1984).

115. Lawrence, "Medicine as Culture," 340.

116. Systems were, to Smith as to Bell, fundamentally beautiful: "That fitness of any system or machine to produce the end for which it was intended, bestows a certain propriety and beauty upon the whole, and renders the very thought and contemplation of it agreeable." Adam Smith, "Of the Effect of Utility upon the Sentiment of Approbation," part IV, in *The Theory of Moral Sentiments* (London: A. Millar, 1759), 79. He adds to that characterization one in which system is methodical, simple, and natural: "The English seem to have employed themselves entirely in inventing, and to have disdained the more inglorious but not less useful labour of arranging and methodizing their discoveries, and of expressing them in the most simple and natural manner. There is . . . no tolerable system of natural philosophy in the English language." Adam Smith, "A Letter to the Editors of the Edinburgh Review, 1756," in Adam Smith and James R. Otteson, *Adam Smith: Selected Philosophical Writings*, Library of Scottish Philosophy (Exeter: Imprint Academic, 2004), 210. For more on this, see also J. R. R. Christie, "Adam Smith's Metaphysics of Language," in *The Figural and the Literal: Problems of Language in the History of Science and Philosophy*, 1630–1800, ed. Geoffrey N. Cantor, Andrew E. Benjamin, and J. R. R. Christie (Manchester: Manchester University Press, 1987).

117. Bell, *Letters*, 171 (March 12, 1810).

118. Dugald Stewart, "Account of the Life and Writings of Adam Smith," in *The Collected Works of Dugald Stewart*, ed. Sir William Hamilton (Edinburgh: Thomas Constable and Co, 1858), 33.

119. Warner, *Against the Spirit of System*.

120. Bell, *Letters*, 118 (December 5, 1807).

121. Philip F. Rehbock, "Transcendental Anatomy," in *Romanticism and the Sciences*, ed. Andrew Cunningham and Nicholas Jardine (Cambridge: Cambridge University Press, 1990); Toby A. Appel, *The Cuvier-Geoffroy Debate: French Biology in the Decades before Darwin* (New York: Oxford University Press, 1987); Robert J. Richards, *The Meaning of Evolution: the Morphological Construction and Ideological Reconstruction of Darwin's Theory*, Science and Its Conceptual Foundations (Chicago: University of Chicago Press, 1992), chap. 3.

122. Among them, the sorts of anatomists whom Desmond describes as conservative, such as London University professor Granville Pattison. Desmond, *Politics of Evolution*, 94–100.

123. For more on the relationship between form and function and on Cuvierian anatomy, see E. S. Russell, *Form and Function: a Contribution to the History of Animal Morphology* (Chicago: University of Chicago Press, 1982); Appel, *Cuvier-Geoffroy Debate*; and Dorinda Outram, *Georges Cuvier: Vocation, Science, and Authority in Post-Revolutionary France* (Manchester: Manchester University Press, 1984).

124. Recall here Bell's passage: "I establish thus a kind of circulation, as it were. In this inquiry I describe many new connections." Bell, *Letters*, 118 (December 5, 1807).

125. Ibid., 170–71 (March 12, 1810).

126. While the antivivisection movement in Britain was largely one born of the second half of the nineteenth century, 1822 legislation (Martin's Act) governing the treatment of animals, as well as public outcry in England over François Magendie's vivisection experiments of roughly the same period, suggest that the British were more reluctant about experimenting on animals. Bell's natural theology also helped to shape his reluctance. For more on vivisection and antivivisectionist movements in the nineteenth century, see Richard D. French, *Antivivisection and Medical Science*

in Victorian Society (Princeton: Princeton University Press, 1975); Anita Guerrini, *Experimenting with Humans and Animals: From Galen to Animal Rights,* Johns Hopkins Introductory Studies in the History of Science (Baltimore: Johns Hopkins University Press, 2003); and Nicolaas A. Rupke, *Vivisection in Historical Perspective* (London: Croom Helm, 1987).

127. Bell, *Letters,* 170–71 (March 12, 1810).

128. Ibid.

129. Ibid.

130. As he wrote to George, "It is this I would print, but the description of the brain I would reserve for more labour of succeeding winters." Ibid., 124 (July 8, 1808).

131. Ibid., 170–71 (March 12, 1810).

132. Ibid., 161 (December 9, 1809).

133. He wrote of his intended contribution and the format he hoped it would take, "[I]n truth, the writing must be short, and yet embracing much . . . It is not meant to explain the anatomy of the Brain, but to state to those who are supposed to know it, the ground and outline of a train of observations, to follow in papers on the Anatomy and Pathology of the Brain, and to establish my claim to these discoveries, if I may yet term them so." Ibid., 125–28 (Summer 1808).

134. Ibid., 125 (July 15, 1808).

135. Ibid., 162 (December 26, 1809).

136. In one example, "As soon as John has transcribed it I'll send it down to you. I expect you will correct it, and have it transcribed, and then give it to Jeffrey and Playfair, as I will to Brougham and some others." Ibid., 125 (July 15, 1808). For other examples, see also letters on pp. 96, 127, and 128.

137. Gordon-Taylor and Walls, *Sir Charles Bell, His Life and Times,* 232.

138. Bell, *Letters,* 125 (July 15, 1808).

139. Ibid., 169 (January 18, 1810).

140. Ibid., 118 (December 5, 1807). In his very first recorded discussion of work on the brain, found, of course, in the context of a letter to his brother, Charles wrote about his plan to lecture on the discovery in order to promote it: "I have done a more interesting nova anatomia cerebri humani than it is possible to conceive. I lectured it yesterday. I prosecuted it last night till one o'clock, and I am sure that it will be well received." Ibid., 117 (November 26, 1807).

141. Ibid., 133 (November 26, 1808).

142. Ibid., 130 (October 17, 1808).

143. Ibid., 133 (November 26, 1808).

144. Ibid., 192 (August 31, 1811).

145. Ibid.

146. Ibid., 159 (November 9, 1809).

CHAPTER TWO

1. Bell, *Letters,* 122 (February 8, 1808).

2. Gordon-Taylor and Walls, *Sir Charles Bell, His Life and Times,* 47–49.

3. Ibid., 49–50.

4. Bell, *Letters,* 203 (7 December 7, 1812).

5. "I am the only lecturer from Scotland. I have gained a high situation in the medical world by my industry; and my character is equal to the situation. I am supporting the school of Edinburgh, and the character of my countrymen as medical men, and the character of the University which produced me. As soon, therefore, as the transaction is completed, I intend to show the nature and pretensions of the school . . . for the support of an old and most respectable institution, one which, founded by the Hunters, has made all the anatomists of the present day, at home and abroad." Ibid., 197 (March 27, 1812).

6. Thomson, "Surgeon-Anatomists of Great Windmill Street School," 56–64.

7. Bell, *Letters*, 199 (May 3, 1812).

8. Ibid., 201 (July 20, 1812).

9. Ibid., 201 (August 25, 1812).

10. Ibid., 208 (October 6, 1813).

11. Ibid., 201–2 (August 25, 1812).

12. William Le Fanu, "Sir Benjamin Brodie," *Notes and Records of the Royal Society of London* 19, no. 1 (1964).

13. Bell, *Letters*, 201–2 (August 25, 1812).

14. Ibid.

15. Ibid.

16. Ibid.

17. William Hunter, *Two Introductory Lectures, Delivered by Dr. William Hunter to his Last Course of Anatomical Lectures, at his Theatre in Great Windmill Street* (London: J. Johnson, 1784), 57.

18. As Bell wrote to his brother George, in Bell, *Letters*, 33 (January 8, 1805).

19. L. S. Jacyna, "Images of John Hunter in the Nineteenth Century," *History of Science* 21 (1983).

20. Alberti, *Morbid Curiosities*; and Alberti and Hallam, *Medical Museums*.

21. Carin Berkowitz, "The Beauty of Anatomy: Visual Displays and Surgical Education in Early Nineteenth-Century London," *Bulletin of the History of Medicine* 85, no. 2 (2011).

22. Bell, *Letters*, 220 (July 1814).

23. This argument about a system of display is discussed in detail in Carin Berkowitz, "Systems of Display: The Making of Anatomical Knowledge in Enlightenment Britain," *British Journal for the History of Science* 46, no. 3 (2013).

24. Matthew Eddy has written about similar themes, though addressing them primarily through looking at the recipients of medical classroom knowledge by examining the visual practices of note-taking: Matthew Eddy, *The Patchwork Picture: Science, Education and the Visual Foundations of Knowledge*, 1760–1820 (Chicago: University of Chicago Press, forthcoming).

25. Hunter, *Two Introductory Lectures*, 6.

26. Ruth Richardson, *Death, Dissection, and the Destitute*, 2nd ed. (Chicago: University of Chicago Press, 2001), 3–30. Richardson's first chapter deals with popular culture and corpses, describing a host of popular rituals surrounding death and corpses (e.g., eating with a dead loved one, taking sacrament with the corpse, and corpse watching). Richardson is concerned with popular and particularly religious meanings and beliefs surrounding corpses, but her discussion makes readily apparent that the physical corpse itself was something of an everyday object. See also Eva Åhrén, *Death, Modernity, and the Body: Sweden* 1870–1940 (Rochester: University of Rochester Press, 2009). For an exploration of the popularity of dissection and anatomy in the American context well into the nineteenth century, see Michael Sappol, *A Traffic of Dead Bodies: Anatomy and Embodied Social Identity in Nineteenth-Century America* (Princeton: Princeton University Press, 2002).

27. John Bell, *Engravings of the Bones, Muscles, and Joints* (London: Longman and Rees, and Cadell and Davies, 1804), xi.

28. Hunter, *Two Introductory Lectures*, 112.

29. Bell, *A System of Operative Surgery*, ii.

30. Hunter, *Two Introductory Lectures*, 109, emphasis in the original. Hunter's use of the phrase "fix it deeper in the mind" is one to which I will return later in the chapter, as it indicates something about what was thought to constitute learning.

31. Bell, *Letters*, 409.

32. E.g., ibid., 199 (April 1812).

33. Charles Bell, "Lectures on the Nervous System Delivered at the College of Surgeons," *London Medical Gazette* 1 (1828): 617.

34. Ibid., 618.

35. Anonymous, "Anonymous Fair-Copy Notes from Lectures on Anatomy and Surgery Reputedly Delivered by Sir Charles Bell in 1822," Brotherton Special Collections, Leeds University Library.

36. See the twenty or so uses of the word "Plan" in John Bell and Charles Bell, *The Anatomy of the Human Body*, 4 vols. (London: T. N. Longman and O. Rees [etc.], 1802). The word is synonymous with "scheme" or "sketch" in John Bell's usage—it implies a general diagram or outline of the anatomy in question.

37. Bell, "Diseases and Accidents to Which the Hip-Joint is Liable," 137.

38. Hunter, *Two Introductory Lectures*, 112.

39. Simon Chaplin, "Nature Dissected, or Dissection Naturalized? The Case of John Hunter's Museum," *Museum & Society* 6 (2008).

40. Hunter, *Two Introductory Lectures*, 55.

41. Anna Maerker, *Model Experts: Wax Anatomies and Enlightenment in Florence and Vienna, 1775–1815* (Manchester: Manchester University Press, 2011).

42. Hunter, *Two Introductory Lectures*, 89.

43. John Hunter had more than seven thousand specimens: Chaplin, "Nature Dissected, or Dissection Naturalized?," 135.

44. "Hunter had many advantages over his rivals. He was a splendid lecturer. He had new anatomical discoveries to impart to his students, and owned better specimens preserved in spirits in glass cases." Roy Porter, "William Hunter, Surgeon," *History Today* 33, no. 9 (1983): 52.

45. Chaplin, "Nature Dissected, or Dissection Naturalized?," 137.

46. Hunter, *Two Introductory Lectures*, 56.

47. Ibid., 57.

48. Bell, *Letters*, 73 (May 19, 1806).

49. Royal College of Surgeons of Edinburgh, Bell Collection, GC 1.43.04

50. William Hunter, Alice Julia Marshall, and John H. Teacher, *Catalogue of the Anatomical Preparations of William Hunter in the Museum of the Anatomy Department* (Glasgow: University of Glasgow, 1970), 661–67. Hunter's catalogue echoes the virtues ascribed to Bell's models, describing one cast as: "A cast in Paris plaster, coloured after life."

51. For more on such museums, see Alberti, *Morbid Curiosities*; Alberti and Hallam, *Medical Museums*; Chaplin, "Nature Dissected, or Dissection Naturalized?"; and Jonathan Reinarz, "The Age of Museum Medicine: The Rise and Fall of the Medical Museum of Birmingham's School of Medicine," *Social History of Medicine* 18, no. 3 (2005).

52. Bell, *Letters*, 202 (October 2, 1812).

53. Bell, *A System of Operative Surgery*, 125.

54. Bell, *Letters*, 200 (June 1, 1812).

55. Royal College of Surgeons of Edinburgh, Bell Collection, BC.xii.2.M.57. GC 11006.

56. For one example, see Hunter, Marshall, and Teacher, *Catalogue of the Anatomical Preparations of William Hunter in the Museum of the Anatomy Department*.

57. Bell, *Letters*, "Lady Bell's Recollections," 410 (Lady Bell's Recollections).

58. Ibid., 176 (June 9, 1810).

59. William Hunter, *Anatomia uteri humani gravidi tabulis illustrata* (Birmingham: John Baskerville, 1774). 3.

60. Charles Bell, *The Hand, Its Mechanism and Vital Endowments as Evincing Design* (London: William Pickering, 1833). See "Unifying Bell's Natural Philosophy and Anatomy through Pedagogy: Understanding Bell's Bridgewater Treatise on the Hand" of this chapter for a more detailed discussion of Bell's treatise and the Bridgewater Treatises more generally.

61. Bell, "Lectures on the Nervous System," 553–54.

62. Ibid.

63. Andrew Cunningham, *The Anatomist Anatomis'd: An Experimental Discipline in Enlightenment Europe* (Farnham: Ashgate, 2010).

64. Charles Bell, *Institutes of Surgery; Arranged in the Order of the Lectures Delivered in the University of Edinburgh*, 2 vols. (Edinburgh: Black, 1838), 1:xx.

65. Ibid.

66. Bell, *The Hand, Its Mechanism and Vital Endowments as Evincing Design*, 329.

67. Ibid., 336.

68. Ibid., 8.

69. Ibid., 192.

70. Bell, *Essays on the Anatomy and Philosophy of Expression*, 7.

71. In addition to objections to the unsavory interactions required to procure bodies for dissection, some within the Royal Academy felt that British artists were displaying anatomy too prominently. According to Martin Kemp, "Knox, writing in 1852 in *Great Artists and Great Anatomists* . . . decried the worst excesses in which 'death-like dissected figures' were displayed on canvas. At its worst, over-exaggerated displays of musculature could become a tired mannerism." Martin Kemp and Marina Wallace, *Spectacular Bodies: The Art and Science of the Human Body from Leonardo to Now* (Berkeley: University of California Press, 2000). 88.

72. Martin Kemp, "True to Their Natures: Sir Joshua Reynolds and Dr. William Hunter at the Royal Academy of Arts," *Notes and Records of the Royal Society of London* 46 (1992); and Harry Mount, "Van Rymsdyk and the Nature-Menders: An Early Victim of the Two Cultures Divide," *British Journal for Eighteenth-Century Studies* 26 (2006).

73. Bell, *Essays on the Anatomy and Philosophy of Expression*, 184.

74. Sarafianos, "B. R. Haydon and Racial Science."

75. "Haydon's teaching was unique in its emphasis on dissection as an essential preparatory step in understanding antique sculpture. Most European academies of art in the late eighteenth century had a lecturer in anatomy, just like the London Royal Academy, but the investigation and study of anatomy by art students was superficial and secondary." Cummings, "B. R. Haydon and His School," 373.

76. Ibid., 370.

77. Bell, *Engravings of the Bones, Muscles, and Joints*, xi.

78. Carin Berkowitz, "Systems of Display: The Making of Anatomical Knowledge in Enlightenment Britain," *British Journal for the History of Science* 46, no. 3 (2013).

79. Charles Bell, *Engravings of the Arteries, Illustrating the Second Volume of the Anatomy of the Human Body* (London: Longman and Rees, 1801). 6.

80. Daston and Galison argue that, as notions of objectivity changed in the nineteenth century, anatomists grew increasingly wary of these ideal types, seeing them as a way for subjectivity to enter their science. As a result, these men began to include depictions of a range of individual, particular bodies in their atlases. Bell, along with his British forebears like Hunter, who themselves regarded their anatomical images as direct depictions of the individual nature on display in front of them, seem to problematize this periodization, or at least to add nuance. For more on idealizing images in atlases, see Lorraine Daston and Peter Galison, *Objectivity* (Cambridge: MIT Press, 2007), 69–83, and "The Image of Objectivity," *Representations* 40 (1992).

81. Bell, *Engravings of the Arteries*, 15.

82. Bell, *Engravings of the Bones, Muscles, and Joints*, 80.

83. The debate is discussed extensively in both Kemp, "True to Their Natures," and Mount, "Van Rymsdyk and the Nature-Menders."

84. Hunter, *Anatomia uteri humani gravidi tabulis illustrata*, preface.

85. Ibid., plate VI.

86. Bell, *Engravings of the Bones, Muscles, and Joints*, vi.

87. H. R., "Some Account of Pestalozzi and His Method of Instruction," *Athenaeum* 2, no. 10 (1807); and Charles Mayo, *A Memoir of Pestalozzi: Being the Substance of a Lecture Delivered at the Royal Institution, Albemarle Street, May, 1826* (London: J. A. Hessey, 1828).

88. For a very good overview of the Scottish Enlightenment, as well as a description of the close-knit circles of philosophically and naturally philosophically minded men who seem to have

produced common themes of inquiry across disciplines and generations, see Broadie, *Cambridge Companion to the Scottish Enlightenment*. Chap. 2 ("Religion and Rational Theology" by M. A. Stewart), chap. 3 ("The Human Mind and its Powers" by Alexander Broadie), and chap. 5 ("Science in the Scottish Enlightenment" by Paul Wood) are particularly instructive on topics covered in this chapter. See also Stuart C. Brown, *British Philosophy and the Age of Enlightenment* (London: Routledge, 1996); and Matthew Eddy, "Converging Paths or Separate Roads? The Roles Played by Science, Medicine, and Philosophy in the Scottish Enlightenment," *Philosophical Writings* 30 (2005).

89. Jonathan Topham, "Science and Popular Education in the 1830s: The Role of the 'Bridgewater Treatises,'" *British Journal for the History of Science* 25, no. 4 (1992); idem, "Beyond the 'Common Context': The Production and Reading of the Bridgewater Treatises," *Isis* 89 (1998); and idem, "Scientific Publishing and the Reading of Science in Nineteenth-Century Britain: A Historiographical Survey and Guide to Sources," *Studies in the History and Philosophy of Science* 31 (2000).

90. For more on natural theology, see Matthew Eddy, "Nineteenth-Century Natural Theology," in *Oxford Handbook of Natural Theology*, ed. Russell Manning (Oxford: Oxford University Press, 2013).

91. Sir Gordon Gordon-Taylor and E. W. Walls set the tone when they proclaimed *The Hand* a natural project for its devout Christian and creationist author. Gordon-Taylor and Walls, *Sir Charles Bell, His Life and Times*, 163–65.

92. Bell, *Letters*, 128 (August 5, 1808).

93. Jennifer Tannoch-Bland, "Dugald Stewart on Intellectual Character," *British Journal for the History of Science* 30 (1997): 311.

94. Stewart, *Elements of the Philosophy of the Human Mind*, vol. 4, part 3, 212.

95. Ibid., 24. Smith's work on the first formation of languages, and its relation to "principles which lead and direct philosophical enquiries" is examined in Christie, "Adam Smith's Metaphysics of Language."

96. Melanie Keene has looked at objects and object lessons in teaching children in her PhD dissertation, "Object Lessons: Sensory Science Education, 1830–1870" (Cambridge: University of Cambridge, 2009). Matthew Eddy has written about the visual practices of note taking; see Eddy, *Seeing the Enlightenment: Learning to Order and Value Visual Culture*.

97. Editor, "XLIX. Proceedings of Learned Societies: the Academy of Sciences at Berlin," *Philosophical Magazine: Comprehending the Various Branches of Science, Liberal and Fine Arts, Agriculture, Manufactures, and Commerce* 18 (1804).

98. H. R., "Some Account of Pestalozzi and His Method of Instruction," 331–32.

99. John Alfred Green, *The Educational Ideas of Pestalozzi* (New York: Greenwood Press, 1969). 220.

100. In that lecture, later printed in its entirety, Mayo said of Pestalozzi's philosophy: "The cultivation of the higher intellectual faculties of reasoning, taste, &c. is preceded by the careful development of just observation and clear intellectual conception. For this purpose, real objects are presented to the examination of the younger pupils; the physical senses are trained to accurate perception, and the understanding is gradually led to generalize and classify the notices it receives through them." Mayo, *A Memoir of Pestalozzi*, 27.

101. Ibid., 16.

102. Bell, *Essays on the Anatomy and Philosophy of Expression*, 16.

103. Bell, *The Hand, Its Mechanism and Vital Endowments as Evincing Design*, 116.

104. Ibid.

105. Stewart argues, "Thus, there is a beautiful and striking analogy among some of the . . . laws of nature; which analogy, however, for anything we know to the contrary, may be the result of the positive appointment of the Creator . . . but, as the evidence of such a connexion does not at least appear satisfactory to every one, it might facilitate the progress of students, and would, at the same time, be fully as unexceptionable in point of sound logic, to establish the fact in particu-

lar cases by experiment and observation, and consider *the law of action and re-action* merely as a general rule or theorem obtained by induction." Stewart, *Elements of the Philosophy of the Human Mind*, vol. 4, part 3, 212–13.

106. Such an assumption supported a kind of inductive science, buttressing a Cuvierian anatomy in which animals were adapted to their environments and each of their anatomical parts was correlated with every other one perfectly. It shored up a proper style of thinking.

107. Bell, *The Hand, Its Mechanism and Vital Endowments as Evincing Design*, 33.

108. Bell, "Lectures on the Nervous System," 553–54.

109. Steven Shapin and Barry Barnes offer useful insight into the relationship between thought and practice, reason and sensation in their article, "Head and Hand: Rhetorical Resources in British Pedagogical Writing, 1770–1850," *Oxford Review of Education* 2, no. 3 (1976). They describe the world Bell inhabited—late Enlightenment Britain—demonstrating the dichotomies posed in pedagogical texts between the reasoned learning of the upper classes and the unthinking sensual and manual learning of the lower orders. Such context is important to understanding the reformist nature of Bell's pedagogy.

110. Bell, *Engravings of the Arteries*, 15–16.

111. Martin Kemp and Susan Lawrence have each mentioned passages in which William Hunter discussed images and memory. Kemp writes: "Hunter's lectures to the Royal Academy of Arts, no less than the preface to the *Gravid Uterus*, show that the highest pleasure to be evoked by a work of art arose when the effects were most truly equivalent to nature herself. The more real the effects, the more the work 'makes stronger impressions on the mind'" (Kemp, "'The Mark of Truth': Looking and Learning in Some Anatomical Illustrations from the Renaissance and the Eighteenth Century," in *Medicine and the Five Senses*, ed. William Bynum and Roy Porter [Cambridge: Cambridge University Press, 1993], 117–18). And Lawrence says, "Hunter and [William] Hamilton both used the central image that being shown the 'object' made the 'impression' formed on the mind by direct observation somehow 'deeper,' hence longer lasting" (Susan Lawrence, "Educating the Senses: Students, Teachers and Medical Rhetoric in Eighteenth-Century London," in Bynum and Porter, *Medicine and the Five Senses*, 170).

CHAPTER THREE

1. Pichot, *Life and Labours of Sir Charles Bell*, 173.

2. Bell, *Letters*, 227–45.

3. Ibid., 269 (January 1, 1821).

4. Ibid., 284 (July 22, 1824).

5. Ibid., 287 (March 16, 1825).

6. Ibid., 288 (March 16, 1825).

7. Ibid., 289 (April 3, 1825).

8. Ibid., 293 (January 9, 1826).

9. Ibid., 214 (March 24, 1814).

10. Ibid., 217 (April 10, 1814).

11. Ibid., 203 (December 7, 1813).

12. Ibid., 299 (January 2, 1828).

13. Roderick Macleod's *London Medical and Physical Journal* and James Johnson's *Medico-Chirurgical Review* were quarterlies that were both in existence before the 1820s, but the bulk of periodicals, and certainly the weeklies, were not founded until later. William Bynum and Janice Wilson, "Periodical Knowledge: Medical Journals and Their Editors in Nineteenth-Century Britain," in *Medical Journals and Medical Knowledge: Historical Essays*, ed. William Bynum, Stephen Lock, and Roy Porter (London: Routledge, 1992), introduction.

14. William Bynum and Janice Wilson write, "Crudely speaking, during the century, London

had 10 per cent of the population and 15 per cent of the doctors. Up to 50 per cent of the profession spent part of their training in the metropolis, almost 75 per cent of the medical journals were published there" ("Periodical Knowledge," 34).

15. Geoffrey Cantor, *Science in the Nineteenth-Century Periodical: Reading the Magazine of Nature* (Cambridge: Cambridge University Press, 2004), 8.

16. Bell, *Institutes of Surgery*, preface.

17. See Bynum and Wilson, "Periodical Knowledge," 38; Michael Brown, "'Bats, Rats and Barristers': *The Lancet*, Libel and the Radical Stylistics of Early Nineteenth-Century English Medicine," *Social History* 39 (2014).

18. Desmond, *Politics of Evolution*, 15–16.

19. Editor, "Medical Journalism—Notice of Parliamentary Inquiry," *London Medical and Surgical Journal* 5 (1834): 55.

20. For similar arguments on the ways in which journals constituted publics, see Thomas Broman, "J. C. Reil and the 'Journalization' of Physiology," in *The Literary Structure of Scientific Argument: Historical Studies*, ed. Peter Dear (Philadelphia: University of Pennsylvania Press, 1991); and idem, "The Habermasian Public Sphere and 'Science in the Enlightenment,'" *History of Science* 36 (1998).

21. See Bell, *Letters*, 23.

22. Samuel Squire Sprigge, *The Life and Times of Thomas Wakley* (London: Longman, Green and Co., 1897), 84–88, 99; Brown, "'Bats Rats and Barristers,'" 189.

23. John Abernethy, *Lectures on the Theory and Practice of Surgery* (London: Longman, Rees, Orme, Brown, and Green, 1830), iii.

24. Ibid.

25. Ibid., iv.

26. Editor, "Cooper v. Wakley," *London Medical Gazette* 3 (1828): 79.

27. Thomas Wakley, "Reply to the Slanderers," *Lancet* 1829–30, vol. 1 (1829): 4.

28. Ibid.

29. Sprigge, *Life and Times of Thomas Wakley*, 105.

30. Ibid., 106.

31. Editor, "Hospital Reporting," *London Medical Gazette* 2 (1828): 120–21.

32. Sprigge, *Life and Times of Thomas Wakley*, 112–13. "'This latter,' said Wakley, 'is so monstrous a proposition that, prepared as we were for the imbecilities of the Hole-and-corner champions, we were somewhat staggered at the impudent absurdity with which it is advanced. Not a scintilla of compassion does the Hole-and-corner advocate suffer to escape him for the victim of the surgeon's want of dexterity; all his sympathy is reserved for the ignorant operator.'"

33. Editor, "Hospital Reporting," *London Medical Gazette* 1 (1828): 697.

34. Ibid., 698.

35. Editor, "Cooper v. Wakley," 83.

36. Wakley, "Reply to the Slanderers," 4–5.

37. See Charles Newman, *The Evolution of Medical Education in the Nineteenth Century* (London, New York: Oxford University Press, 1957); Irvine Loudon, "Medical Practitioners 1750–1850 and the Period of Medical Reform in Britain," in *Medicine in Society*, ed. Andrew Wear (Cambridge: Cambridge University Press, 1992).

38. See John Harley Warner, "The Idea of Science in English Medicine: The 'Decline of Science' and the Rhetoric of Reform, 1815–1845," in *British Medicine in an Age of Reform*, ed. Roger French and Andrew Wear (New York: Routledge, 1991), 138; and Russell Maulitz, "Channel Crossing: the Lure of French Pathology for English Medical Students, 1816–1836," *Bulletin of the History of Medicine* 55 (1981).

39. For more on French medicine in the early nineteenth century, see Erwin Ackerknecht, *Medicine at the Paris Hospital, 1794–1848* (Baltimore: Johns Hopkins University Press, 1967); W. F. Bynum, *Science and the Practice of Medicine in the Nineteenth Century* (Cambridge: Cambridge

University Press, 1994), chapters 2 and 4; and Brockliss and Jones, *Medical World of Early Modern France*. For two studies of the development of French medicine in the eighteenth century, both of which find that there was much going on in a period often seen as being stagnant, see L. W. B. Brockliss, "Before the Clinic: French Medical Teaching in the Eighteenth Century," in *Constructing Paris Medicine*, ed. Caroline Hannaway and Ann La Berge (Amsterdam: Rodopi, 1998); and Gelfand, *Professionalizing Modern Medicine*.

40. *London Medical and Surgical Journal* (1832), 1:109–10.

41. Ian Burney offers a succinct and helpful summary of the elements that British medical men admired in French medicine, beginning with structural differences between systems and ending with the localization of disease in the lesion, which was to be identified during routinized autopsy. Ian Burney, "Medicine in the Age of Reform," in *Rethinking the Age of Reform: Britain 1780–1850*, ed. Arthur Burns and Joanna Innes (Cambridge: Cambridge University Press, 2003), 167.

42. See Mary Bostetter, "The Journalism of Thomas Wakley," in *Innovators and Preachers: The Role of the Editor in Victorian England*, ed. Joel H. Wiener (Westport: Greenwood, 1985); Warner, "Idea of Science in English Medicine," 142; and Keir Waddington, *Medical Education at St. Bartholomew's Hospital, 1123–1995* (Woodbridge: Boydell Press, 2003). 142.

43. Anonymous ("Argus"), "Letter to the Editor: Medical and Surgical Reform," *Lancet* 1828–29, vol. 2 (1828): 397.

44. The Royal College of Surgeons had been known for its corruption and financial mishaps since the eighteenth century. See Hamilton, "Medical Professions in the 18th Century."

45. Anonymous ("Juvenis"), "Apothecaries," *Lancet* 1828–29, vol. 2 (1828): 429.

46. Loudon, "Medical Practitioners 1750–1850 and the Period of Medical Reform in Britain."

47. Thomas Wakley, "Editorial," *Lancet* 1829–30, vol. 1 (1829): 42–43.

48. Bonner quotes one student who wrote to the *Lancet* in 1826, "The medical lectures [. . .] were 'written compositions read over to the students.' To be of value, he said such lectures had to be made 'as *clinical* as possible.' Another student complained as late as 1842 that although he had been in London for six weeks, he had not heard a single clinical lecture." Thomas Neville Bonner, *Becoming a Physician: Medical Education in Britain, France, Germany, and the United States, 1750–1945* (New York: Oxford University Press, 1995), 133.

49. Desmond, *Politics of Evolution*, 104.

50. Editor, "Criminal Information against the Rioters—New Bye-Laws of the College of Surgeons," *London Medical Gazette* 8 (1831): 279.

51. The utility and drawbacks of looking at the history of science through the lens of "national styles" have already been revealed; see Gerald Geison and Frederic Holmes, eds., *Research Schools: Historical Reappraisals, Osiris* 8 (Chicago: University of Chicago Press, 1993), 30–49; Jonathan Harwood, *Styles of Scientific Thought: The German Genetics Community, 1900–1933*, Science and Its Conceptual Foundations (Chicago: University of Chicago Press, 1993). Though the rhetoric of the conservative reformers looks as though it would fit some such arguments, I am not interested in engaging in an analysis of whether there really were national styles of medical science. There were, of course, differences in the kinds of educational institutions used in different countries that helped to shape their medical sciences, but I am centrally concerned with looking at why the conservative reformers employed a rhetoric that emphasized national differences, real or perceived.

52. Burns and Innes, *Rethinking the Age of Reform: Britain 1780–1850*, introduction.

53. It is particularly reminiscent of writings of Edmund Burke and those who followed, adopting his stances on revolution, reform, and British exceptionalism; see Burns and Innes, *Rethinking the Age of Reform: Britain 1780–1850*, 88; and Burney, "Medicine in the Age of Reform." Burke's stance shows just how piecemeal British reform philosophies could be—most Whigs endorsed parliamentary reform, but clearly Burke did not. That even Burke could be called a reformer shows that most platforms during the Age of Reform involved at least some reformist rhetoric—almost no one was arguing for things to remain exactly as they were. For an example of Burkean rhetoric, see, e.g., Editor, "Reform—College of Physicians," *London Medical Gazette* 11 (1832–33).

54. Burns and Innes, *Rethinking the Age of Reform: Britain 1780–1850*, 14–15.

55. Editor, "Medical Education," *London Medical Gazette* 1 (1828): 314.

56. Editor, "Medical Reform—Education," *London Medical Gazette* 11 (1832): 90.

57. Nationalism has been the subject of several recent books on Britain, though most do not engage with science or medicine. Conservative reform arguments suggest that the idea of the nation was simultaneously being built through debates over medical reform. For more on the creation of British nationalism, see Linda Colley, *Britons: Forging the Nation, 1707–1837* (New Haven: Yale University Press, 1992); Colin Kidd, *British Identities Before Nationalism: Ethnicity and Nationhood in the Atlantic world, 1600–1800* (Cambridge: Cambridge University Press, 1999); Eric J. Evans, *The Forging of the Modern State: Early Industrial Britain, 1783–1870*, Foundations of Modern Britain (London: Longman, 1983); and E. J. Hobsbawm and T. O. Ranger, *The Invention of Tradition* (Cambridge: Cambridge University Press, 1983).

58. Porter, "Medical Lecturing in Georgian London," 96.

59. Editor, "Present State of the London and Paris Schools of Medicine," *London Medical Gazette* 7 (1830): 24.

60. See chap. 4 of this book for more on London University.

61. Charles Bell, "London University—Mr. Bell's Introductory Lecture," *London Medical Gazette* 5 (1830): 18.

62. Rosner, *Medical Education in the Age of Improvement*, chap. 3; and Christopher Lawrence, "The Shaping of Things to Come: Scottish Medical Education 1700–1939" *Medical Education* 40, no. 3 (2006).

63. Bynum, *Science and the Practice of Medicine*, 4.

64. For example, Roy Porter writes, "Such courses had many attractions. Some embodied discoveries unavailable elsewhere. William Hunter's auditors heard of his researches on aneurysm, the placental circulation, the lymphatic system and the gravid uterus, nowhere available in print" ("Medical Lecturing in Georgian London," 94).

65. Bell, *Letters*, 159 (November 9, 1809).

66. Charles Bell, "Clinical Lecture on Diseases of the Spine, Delivered, Nov. 3, at the Middlesex Hospital School," *London Medical Gazette* 17 (1835): 231.

67. Editor, "London University—Apprenticeships," *London Medical Gazette* 8 (1831): 336.

68. Bonner, *Becoming a Physician*, 139.

69. Anonymous ("Voyageur"), "Letter to the Editor: Parisian Medicine," *London Medical Gazette* 1 (1828): 695.

70. As was the case with William Sharpey's testimony about Magendie's experiments before the Royal Commission on Vivisections: "In the first place, they were painful (in those days there were no anaesthetics), and sometimes they were severe; and then THEY WERE WITHOUT SUFFICIENT OBJECT. For example, Magendie made incisions into the skin of rabbits and other creatures TO SHOW THAT THE SKIN IS SENSITIVE! Surely all the world knows the skin is sensitive; no experiment is wanted to prove that." Great Britain Royal Commission on Vivisection, *Report of the Royal Commission on the Practice of Subjecting Live Animals to Experiments for Scientific Purposes* (London: George Edward Eyre and William Spottiswoode, 1876), 22, question 444.

71. James Macauley to Lyon Playfair, May 13, 1875 (Imperial College of Science and Technology, Playfair Papers #469), quoted in Richard D. French, *Antivivisection and Medical Science in Victorian Society* (Princeton: Princeton University Press, 1975): 276–79.

72. Anonymous, "Charles Bell's Bridgewater Treatise," *London Medical Gazette* 13 (1833): 254. "Torturing nature" was, of course, also a Baconian trope popularized in this period by William Whewell.

73. This advertisement was taken from Bell, *Institutes of Surgery*.

74. Benjamin Brodie, "Introductory Discourse on the Studies Required for the Medical Profession," in *The Works of Sir Benjamin Collins Brodie*, ed. Charles Hawkins (London: Longman, Green, Longman, Roberts, and Green, 1838), 467.

75. Ibid.

76. Richard Hunt and Ida Macalpine, "A Privately Printed Pamphlet by Sir Charles Bell on the Principles Involved in Appointments to the London Hospitals," *Annals of the Royal College of Surgeons of England* 30 (1962).

77. Ibid., 261. They write of the pamphlet, "As Bell has not so far been credited with an active interest in reforms of the medical profession, we tried to find a copy of his pamphlet in the London libraries to study it, but were unsuccessful [. . .] at last we found it listed in W. B. Taylor's *Catalogue of the Library of the Society of Apothecaries*, 1913, listed anonymously under Middlesex Hospital as number 4 of 13 pamphlets" (258). Copies of this catalogue had been auctioned off and they were able to get hold of one only through correspondence with a collector, so their article remains the only accessible source of this piece by Bell.

78. Ibid.

79. Quoted in Porter, "Medical Lecturing in Georgian London," 93.

80. Editor, "Professional Sketches," *Medical Times* 3, no. 57 (1840): 37.

81. Thomas Wakley, "London College of Medicine: May 7, 1831," *Lancet* 1830–31, vol. 2 (1831). This meeting helps to substantiate my claim about the social and political coherence of radical reformers, who composed a small enough group that they met and established agreement and leadership. Conservative reformers never participated in such political gatherings meant to develop platforms for reform.

82. Ibid.

83. For example, Wakley wrote, "We have carefully examined the machinery of medical Colleges and Universities in this country and throughout the world" in establishing the London College of Medicine. Anonymous, "London College of Medicine, Second Public Meeting of the Supporters of this Institution," *Lancet* 1830–31, vol. 2 (1831): 217.

84. Wakley, "London College of Medicine: May 7, 1831," 181.

85. "Will young men, freed from the restraints imposed by the 'ticket system,' as it has been called, and left to their own pleasure and discretion in the attendance upon, and acquisition of means of improvement, obtain the same information, acquire as much practical knowledge as under the present system?" Anonymous ("Omega"), "Letter to the Editor: London College of Medicine," *Lancet* 1830–31, vol. 2 (1831): 345.

86. L., "London College of Medicine: Objections to the Universal Title of 'Doctor,'" *Lancet* 1830–31, vol. 2 (1831): 687.

87. Anonymous, "Letter to the Editor: London College of Medicine," *Lancet* 1830–31, vol. 2 (1831): 256.

88. Anonymous, "Letter to the Editor: Collegium Wakleyanum," *London Medical Gazette* 8 (1831): 272.

89. Ibid.

90. Editor, "College of Surgeons," *London Medical Gazette* 7 (1831): 789.

91. Anonymous ("One of the Multitude"), "College of Surgeons," *London Medical Gazette* 7 (1831): 820

92. Joseph Henry Green, "Distinction Without Separation," *London Medical Gazette* 8 (1831): 217.

93. Anonymous, "Letter to the Editor: Collegium Wakleyanum," 272.

94. Ibid., 271.

95. Editor, "College of Surgeons," 790.

96. Thomas Wakley, "London College of Medicine: June 18, 1831," *Lancet* 1830–31, vol. 2 (1831): 379.

97. Thomas Wakley, "London College of Medicine," *Lancet* 1830–31, vol. 2 (1831): 244.

98. Anonymous, ("Philomeides"), "Letter to the Editor: Grand Convocation of the Collegium Wakleyanum," *London Medical Gazette* 8 (1831): 208.

99. Richardson, *Death, Dissection, and the Destitute*, 42.

100. Iwan Rhys Morus, "Radicals, Romantics and Electrical Showmen: Placing Galvanism at

the End of the English Enlightenment," *Notes and Records of the Royal Society* 63 (2009): 268. See also Richardson, *British Romanticism and Science of the Mind*, 117.

101. Bransby Blake Cooper, *The Life of Sir Astley Cooper, Bart., Interspersed with Sketches from His Note-Books of Distinguished Contemporary Characters* (London: J. W. Parker, 1843), 225.

102. Both Benjamin Brodie and Astley Cooper were well-known teachers and outspoken advocates for reform to the anatomy laws. Brodie advocated operating less, particularly doing fewer amputations, and treating surgical cases medically. He also conducted vivisection experiments. Benjamin Brodie, *Pathological and Surgical Observations on Diseases of the Joints* (London: Longman, Hurst, Rees, Orme, and Brown, 1818). Brodie had served as a demonstrator at the Great Windmill Street School of Anatomy under James Wilson, from whom Bell bought the school. He was elected to the Council of the Royal College of Surgeons in 1829, despite having "signed a memorial to the Council in 1826, suggesting the need to reform its constitution." Fanu, "Sir Benjamin Brodie," 48. In 1840 he helped to pass a reform to make the council elective from all of the fellows.

103. Penelope Hunting, *History of the Royal Society of Medicine* (London: Royal Society of Medicine, 2001), chap. 3.

CHAPTER FOUR

1. Bell, *Letters*, 343 (7 October 7, 1835).

2. Anonymous, "Museum Making," *Lancet* 1826–27, vol. 2 (1827): 374.

3. Gordon-Taylor and Walls, *Sir Charles Bell, His Life and Times*.

4. Bell, *Letters*, 281 (December 28, 1823).

5. Ibid., 294 (16 February 16, 1826).

6. Ibid., 295 (19 January 19, 1827).

7. Lawrence, "Entrepreneurs and Private Enterprise."

8. Students would often pay to be allowed to walk through a hospital as someone's pupil, gaining access to sick patients and sometimes to the clinical notes of the doctors and surgeons. Susan Lawrence has written extensively on both the private schools of anatomy and on ward walking and hospital education in Lawrence, "Entrepreneurs and Private Enterprise," and *Charitable Knowledge*.

9. Bonner, *Becoming a Physician*, 172–73; and Desmond, *Politics of Evolution*, 31–32, 92–94.

10. Bell, *Letters*, 300.

11. Charles Bell, "Sir Charles Bell's Introductory Address on the Opening of London University," *London Medical Gazette* 2 (1828): 566.

12. Ibid.

13. *Statement by the Council of the University of London, Explanatory of the Nature and Objects of the Institution* (London: Longman, Rees, Orme, Brown, and Green; and John Murray, Albemarle Street, 1827), 8.

14. Ibid., 9. "There are now 6,000 members of the College of Surgeons, not six of whom, it has been stated, have graduated at the universities. In the higher branch of law, a very considerable proportion have graduated at Oxford and Cambridge; but among those, who belong to a very important branch of the profession—the attornies [*sic*], of whom there are not less than eight thousand in England, it is believed that scarcely one in a thousand has had the advantages of an university education. Those, who hold places in the offices of government, a class that ought to enjoy the benefits of a liberal education, are also unable to avail themselves of the facilities afforded at Oxford and Cambridge, because they usually enter such offices at or before the age of the youngest under-graduate of those universities."

15. Thomas Babington Macaulay, "Thoughts on the Advancement of Academical Education in England," *Edinburgh Review* 44 (1826): 324.

16. Ibid.

17. Geoffrey Cantor, "Henry Brougham and the Scottish Methodological Tradition," *Studies in History and Philosophy of Science* 2, no. 1 (1971).

18. N. B. Harte, *The University of London, 1836–1986: An Illustrated History* (London: Athlone Press, 1986), 14.

19. Edmund Burke, "The London University," *Annual Register* 67 (1826).

20. Ibid.

21. *Statement by the Council of the University of London, Explanatory of the Nature and Objects of the Institution*, 19.

22. Ibid., 31.

23. Note that medicine and surgery were granted as a unified degree, a rare mixing of professions that traditionally had been separated by prestige, as well as institution, in London.

24. Leonard Horner, "University of London, Medical Diploma," *London Medical Gazette* 6 (1830), 219–21.

25. Bell, *Letters*, 299–300 (January 2, 1828).

26. Irvine Loudon, *Medical Care and the General Practitioner, 1750–1850* (Oxford: Clarendon Press, 1986).

27. Jacyna, "Images of John Hunter in the Nineteenth Century."

28. For more on the Apothecaries' Act and general practitioners, see Loudon, *Medical Care and the General Practitioner, 1750–1850*, chaps. 7–8, pp. 151–89. See also Lawrence, "Private Enterprise and Public Interests"; Charles Singer and S. W. F. Holloway, "Early Medical Education in England in Relation to the Pre-History of London University," *Medical History* 4, no. 1 (1960); S. W. F. Holloway, "The Apothecaries Act," *Medical History* 10, no. 3 (1966); and Loudon, *Medical Care and the General Practitioner, 1750–1850*, and "Medical Practitioners 1750–1850 and the Period of Medical Reform in Britain."

29. Lawrence, "Entrepreneurs and Private Enterprise."

30. Horner, "University of London, Medical Diploma."

31. Bell, *Letters*, 300 (June 1828).

32. Ibid., 252 (December 15, 1815).

33. Bell, *Institutes of Surgery*, 1:xix–xxii.

34. John Cohen, "Medical Education in the University of London, University College and Middlesex Hospitals 1800–1840" (MA thesis, University of London, 1991).

35. Ibid., 68.

36. According to the *London Magazine*, the following were members of the founding council of the University: Hon. James Abercrombie, MP; Zachary Macauley, Esq., FRS; Right Hon. Lord Auckland; Sir James Mackintosh, MP, FRS; Alexander Baring, Esq., MP; James Hill, Esq.; George Birkbeck, MD; His Grace the Duke of Norfolk; Henry Brougham, Esq., MP, FRS; Lord John Russell, MP; Thomas Campbell, Esq.; Benjamin Shaw, Esq.; Right Hon. Viscount Dudley and Ward; John Smith, Esq., MP; Isaac Lyon Goldsmid, Esq.; William Tooke, Esq., FRS; O. G. Gregory, LL.D; Henry Warburton, Esq., FRS; George Grote, Jun. Esq.; Henry Weymouth, Esq.; Joseph Hume, Esq., MP, FRS; John Wishaw, Esq., FRS; Most Noble the Marquis of Lansdowne, FRS; and Thomas Wilson, Esq. "Editorial: London University," *London Magazine* 5 (1826): 554.

37. "Editorial: London University—Mr. Bell," *London Medical Gazette* 7 (1830): 309.

38. See Cohen, "Medical Education," 82: "The dispute was originally about teaching [. . .] but it became involved in issues of procedure, the organization, administration and constitutional structure of the University. The true situation was, however, complicated by personal enmities and bitterness, and the press was bombarded by letters and pamphlets from the various parties."

39. Editor, "Memorial of the Medical Teachers," *London Medical Gazette* 14 (1834).

40. *Statement by the Council of the University of London, Explanatory of the Nature and Objects of the Institution*, 15.

41. Cohen, "Medical Education," 65.

42. "Editorial: London University—Mr. Bell."

43. Bell, *Letters*, 311 (March 1, 1830).

44. "Editorial: Resignation of Mr. C. Bell," *London Medical Gazette* 6 (1830): 473.

45. For more on the significance of instrumentality, the ability to *do something*, in modern science, see Peter Dear, *The Intelligibility of Nature: How Science Makes Sense of the World* (Chicago: University of Chicago Press, 2006), 173–95. This sort of instrumentality seems to resonate with the sort of medical science being created by conservative reformers.

46. Editor, "Memorial of the Medical Teachers," 242.

47. See chap. 3 of this book.

48. Charles Bell, "Mr. Bell's Letter to his Pupils of the London University, on Taking Leave of Them," *London Medical Gazette* 7 (1830): 308.

49. Editor, "St. Bartholomew's Hospital," *Lancet* 1829–30, vol. 1 (1829): 47. Emphasis in the original.

50. John Elliotson, "Introductory Address to a Course of Lectures on the Principles and Practice of Medicine," *Lancet* 1831–32, vol. 1 (1831): 64–65.

51. Ibid., 65.

52. William Coleman and Frederic Lawrence Holmes, *The Investigative Enterprise: Experimental Physiology in Nineteenth-Century Medicine* (Berkeley: University of California Press, 1988); and Lesch, *Science and Medicine in France.*

53. For more on Bennett and on the scandal surrounding Pattison's dismissal, see Desmond, *Politics of Evolution*, 81–100.

54. Bell, *Letters*, 318 (January 1831).

55. The passage ends with sentences quoted earlier in this chapter: "for it is impossible that medicine, as a practical science, can be taught without a constant reference to the chambers of the sick, any more than chemistry can be taught without apparatus, botany without plants, or anatomy without bodies." "Editorial: Resignation of Mr. C. Bell," 473.

56. Desmond, *Politics of Evolution*, chap. 2.

57. Lawrence, "Entrepreneurs and Private Enterprise" and *Charitable Knowledge*, chap. 4.

58. Susan Lawrence has written, "The Guy's 1807 announcement that the hospital lectures, together with courses at St. Thomas's, formed a 'school' illustrates the gradual development of a collective identity, a moving away from the situation characterized by the individual advertisements." In Lawrence, "Entrepreneurs and Private Enterprise," 182.

59. Waddington, *Medical Education at St. Bartholomew's Hospital*, 38.

60. Ibid.

61. The address was signed by the physicians Francis Hawkins and Thomas Watson, and by the surgeons Charles Bell, Herbert Mayo, and Edward Tuson. Gordon-Taylor and Walls, *Sir Charles Bell, His Life and Times*, 148.

62. Reproduced in William James Erasmus Wilson, *The History of the Middlesex Hospital During the First Century of its Existence* (Oxford: Oxford University, 1843), 165.

63. Ibid., 169.

64. Ibid., 168.

65. Charles Bell, "Middlesex Hospital School: Opening Address," *Lancet* 1835–36, vol. 1 (1835): 89.

66. Ibid.

67. Ibid.

68. Bell, *Letters*, 341 (June 29, 1835).

69. Waddington, *Medical Education at St. Bartholomew's Hospital, 1123–1995*, 36.

70. Bell, "Middlesex Hospital School: Opening Address," 89.

71. Bell, *Institutes of Surgery*, 1:xix.

72. Michel Foucault, *The Birth of the Clinic: An Archaeology of Medical Perception* (New York: Pantheon Books, 1973); Brockliss, "Before the Clinic"; and Coleman and Holmes, *Investigative Enterprise.*

73. Bynum, *Science and the Practice of Medicine*, 47–49.

74. See chap. 3 of this book for examples of such juxtapositions. One student, for example, quoted on page 108, wrote, "Indeed, they [the French] seem to think that the perfection of medicine consists not so much in keeping patients alive as in foretelling with precision the appearances which will be found after death." Anonymous ("Voyageur"), "Letter to the Editor: Parisian Medicine," 695.

75. Brodie, "Introductory Discourse on the Studies Required for the Medical Profession," 470.

76. Gordon-Taylor and Walls, *Sir Charles Bell, His Life and Times*, 149.

77. Bell, "Clinical Lecture on Diseases of the Spine, Delivered, Nov. 3, at the Middlesex Hospital School."

78. Ibid., 231.

79. For example, Bell wrote, "Your long list of certificates you must have; but I conjure you to act as if anatomy, and such uses of anatomy as you see in hospital practice, were the business of your life in London, and not to be satisfied with learning to answer such questions as may be put to you at any board" (ibid.).

80. Editor, "Introductory Address on the Opening of the Middlesex Hospital Medical School," *Lancet* 1835–36, vol. 1 (1835): 148.

81. Ibid. Emphasis in the original.

82. Bell, *Letters*, 345 (November 27, 1835).

83. Waddington, *Medical Education at St. Bartholomew's Hospital, 1123–1995*, 68.

84. Brodie, "Introductory Discourse on the Studies Required for the Medical Profession," 469.

85. Ibid., 6–7.

86. Bell, *Institutes of Surgery*, 1:xx.

87. Brodie, "Introductory Discourse on the Studies Required for the Medical Profession," 468.

88. Ibid.

89. Bell, "Clinical Lecture on Diseases of the Spine, Delivered, Nov. 3, at the Middlesex Hospital School."

90. Bell, *Institutes of Surgery*, 1:xxi–xxii.

91. Brodie, "Introductory Discourse on the Studies Required for the Medical Profession," 467.

92. See Desmond, *Politics of Evolution*, chap.2, "Importing the New Morphology."

93. Ibid., 82.

94. Waddington, *Medical Education at St. Bartholomew's Hospital, 1123–1995*, 75.

95. Gerald L. Geison, *Michael Foster and the Cambridge School of Physiology: The Scientific Enterprise in Late Victorian Society* (Princeton: Princeton University Press, 1978), 329; and idem, "Divided We Stand: Physiologists and Clinicians in the American Context," in *The Therapeutic Revolution: Essays in the Social History of American Medicine*, ed. Morris J. Vogel and Charles Rosenberg (Philadelphia: University of Pennsylvania Press, 1979); Samuel Shortt, "Physicians, Science, and Status: Issues in the Professionalization of Anglo-American Medicine in the Nineteenth Century," *Medical History* 27 (1983); and Harry M. Marks, *The Progress of Experiment: Science and Therapeutic Reform in the United States*, 1900–1990, Cambridge History of Medicine (Cambridge: Cambridge University Press, 1997).

96. Bell, *Letters*, 345 (November 27, 1835).

97. For more on British ideas about experimentation, see chap. 2 and Carin Berkowitz, "Disputed Discovery: Vivisection and Experiment in the 19th Century," *Endeavour* 30, no. 3 (2006).

98. Bell, "Sir Charles Bell's Introductory Address on the Opening of London University," 568.

CHAPTER FIVE

1. Bell, *Letters*, 338.

2. Ibid., 347 (December 8, 1835).

3. Ibid., 358 (September 3, 1837).

4. Charles Bell to Henry Brougham, January 10 [year not noted], #24806, Brougham Papers, Special Collections, University College London.

5. Bell, *Letters*, 359 (May 15, 1838).

6. Ibid., 366 (January 17, 1839).

7. Charles Bell to Henry Brougham, November 19 [year not noted], #24809, Brougham Papers, Special Collections, University College London.

8. Bell, *Letters*, 355 (July 8, 1837).

9. Alexander Shaw, *Narrative of the Discoveries of Sir Charles Bell in the Nervous System* (London: Longman, 1839).

10. A version of this chapter, addressing the disputed discovery of the roots of motor and sensory nerves, appears as Carin Berkowitz, "Defining a Discovery: Priority and Methodological Controversy in Early Nineteenth-Century Anatomy," *Notes and Records of the Royal Society* 68, no. 4 (2014), in press.

11. Charles Bell, *Idea of a New Anatomy of the Brain: A Facsimile of the Privately Printed Edition of* 1811, *with a Bio-Bibliographical Introduction* (London: Dawsons of Pall Mall, 1966). See also chap.1.

12. Ibid., 21.

13. In a letter to his brother written in 1807, Bell says "I establish thus a kind of circulation, as it were. In this inquiry I describe many new connections. The whole opens up in a new and simple light; the nerves take a simple arrangement; the parts have appropriate nerves; and the whole accords with the phenomena of the pathology, and is supported by interesting views once in wisdom; not pieced together like the work of human ingenuity": Bell, *Letters*, 117–18 (December 5, 1807). He later writes that Magendie experiments "in hope to catch at some of the accidental facts of a system which, it is evident, the experimenters did not fully comprehend": Charles Bell, *An Exposition of the Natural System of the Nerves of the Human Body* (London: Spottiswode, 1825), 3–4.

14. See chap. 1.

15. François Magendie, "Expériences sur les fonctions des racines des nerfs rachidiens," *Journal de physiologie expérimentale et de pathologie* 2 (1822), and "Expériences sur les fonctions des racines des nerfs qui naissent de la moëlle épinière," *Journal de physiologie expérimentale et de pathologie* 2 (1822).

16. T. C. Hansard, "Hansard's Parliamentary Reports, February 24" (London, 1825), 657.

17. Ibid.

18. As judged in Olmsted, *François Magendie, Pioneer in Experimental Physiology*, 141.

19. French, *Antivivisection and Medical Science*, 18–21.

20. Bell, *Letters*, 323 (October 12, 1831).

21. There is a secondary literature on issues of priority in science that, in its more theoretical guise, arises chiefly from Robert Merton's work in the 1950s (e.g., Robert K. Merton, "Priorities in Scientific Discovery: A Chapter in the Sociology of Science," *American Sociological Review* 22, no. 6 [1957]), in which Merton argued for the functional value of priority attribution in science and attempted thereby to explain the occurrence of disputes over the matter; most subsequent literature on priority issues has focused on the specifics involved in particular disputes, where the significance therefore lies in matters distinct from the general category "priority dispute" itself (such as the Newton-Leibniz dispute over the calculus); see Augustine Brannigan, *The Social Basis of Scientific Discoveries* (Cambridge: Cambridge University Press, 1981). Perhaps the most relevant piece of recent years is David Philip Miller, *Discovering Water: James Watt, Henry Cavendish and the Nineteenth Century "Water Controversy"* (Aldershot: Ashgate, 2004), on the after-the-fact dispute over who should receive credit for determining the compound character of water, a dispute that involved some of the same figures as feature in this book (such as Henry Brougham and Francis Jeffrey)—though that too is a story of a particular dispute and not of priority disputes in general, and attempts to understand what was at stake for various actors in their claims to priority.

22. Bell, *Letters*, 96 (May 21, 1807), 117 (November 31, 1807). This eighteenth-century discovery was famous because it too produced a priority dispute, this one between William Hunter and Alexander Monro Secundus.

23. Ibid., 259 (March 2, 1818).

24. From a letter to his brother, quoted in Pichot, *Life and Labours of Sir Charles Bell*, 84.

25. Bell, *Idea of a New Anatomy of the Brain.*

26. See, e.g., Bell, *Idea of a New Anatomy of the Brain*; and Sir Charles Bell, "Clinical Lecture on Diseases of the Nerves," *London Medical Gazette* 13 (1833–34).

27. For more on Gall's influence on nineteenth-century work on the nerves and brain, see Shapin, "Phrenological Knowledge and the Social Structure of Early Nineteenth-Century Edinburgh"; and Edwin Clarke and L. S. Jacyna, *Nineteenth-Century Origins of Neuroscientific Concepts* (Berkeley: University of California Press, 1987), introduction, for more on phrenology in nineteenth-century Britain.

28. The quintessential nineteenth-century work on natural theology was William Paley, *Natural Theology* (London: printed for R. Faulder by Wilks and Taylor, 1802). Bell even edited a version in 1835. He also wrote one of the Bridgewater Treatises, a series of works attesting to the "Power, Wisdom, and Goodness of God, as manifested in the Creation": Bell, *The Hand, Its Mechanism and Vital Endowments as Evincing Design*. For an analysis of natural theology by historians, see Aileen Fyfe, "Publishing and the Classics: Paley's Natural Theology and the Nineteenth-Century Scientific Canon," *Studies in History and Philosophy of Science* 33 (2002); and L. S. Jacyna, "Immanence or Transcendence: Theories of Life and Organization in Britain 1790–1835," *Isis* 74 (1983).

29. Bell, *Idea of a New Anatomy of the Brain*, 7.

30. Ibid., 5–6.

31. Bell, *Letters*, 171.

32. Charles Bell, "Idea of a New Anatomy of the Brain, with Letters &c.," *Journal of Anatomy and Physiology* 3, no. 1 (1868): 154.

33. Ibid., 153.

34. Bell, *Idea of a New Anatomy of the Brain*, 4–6.

35. Bell, *Letters*, 117–18 (December 5, 1807).

36. Bell, *Idea of a New Anatomy of the Brain*, 15.

37. Ibid., 20.

38. Shapin, "Phrenological Knowledge and the Social Structure of Early Nineteenth-Century Edinburgh."

39. James A. Harris, *Of Liberty and Necessity: The Free Will Debate in Eighteenth-Century British Philosophy* (Oxford: Oxford University Press, 2005).

40. Bell, *Letters*, 170 (March 12, 1810).

41. Bell wrote, "[t]o this end I made experiments which, though they were not conclusive, encouraged me in the view I had taken." Bell, *Idea of a New Anatomy of the Brain*, 21–22.

42. Ibid., 33.

43. Bell, *Letters*, 118 (December 5, 1807).

44. Stephen Jacyna, "Theory of Medicine, Science of Life: The Place of Physiology Teaching in the Edinburgh Medical Curriculum, 1790–1870," in *The History of Medical Education in Britain*, ed. Vivian Nutton and Roy Porter (Amsterdam: Rodopi BV Editions, 1995).

45. Bell, *Letters*, 275 (June 10, 1822).

46. For more on changes in print culture, see Broman, "J. C. Reil and the 'Journalization' of Physiology"; James A. Secord, *Victorian Sensation: The Extraordinary Publication, Reception, and Secret Authorship of Vestiges of the Natural History of Creation* (Chicago: University of Chicago Press, 2000); and Alex Csiszar, "Seriality and the Search for Order: Scientific Print and its Problems during the Late Nineteenth Century," *History of Science* 48, no. 3/4 (2010).

47. See Cranefield and Bell, *The Way In and the Way Out*; C. B. Jorgensen, "Aspects of the History of the Nerves: Bell's Theory, the Bell-Magendie Law and Controversy, and Two Forgotten Works by P. W. Lund and D. F. Eschricht," *Journal of the History of Neuroscience* 12, no. 3 (2003); and Olmsted, *François Magendie, Pioneer in Experimental Physiology*.

48. According to Clarke and Jacyna (*Nineteenth-Century Origins of Neuroscientific Concepts*), "It

is now widely accepted that although Bell made the first experimental observations on spinal root properties, his claims for full priority cannot be allowed, for two reasons. First, his pioneer, but sole, investigation was incomplete and the results he obtained did not warrant the conclusions deduced, which in any case were mainly erroneous. Second, and of much more sinister significance, is the damning evidence against Bell that, in an attempt to establish his leadership, he dishonestly appropriated Magendie's correct opinions and in light of them deceitfully emended his earlier publications before reprinting them in order to support his case" (111).

49. For other accounts of the priority dispute, see Pierre Flourens, "1858 Memoir of Magendie, Translated by C. A. Alexander," *Annual Report of the Smithsonian Institution* (1866); Lesch, *Science and Medicine in France*; Pauline Mazumdar, "Anatomy, Physiology and Surgery: Physiology Teaching in Early Nineteenth-Century London," *Canadian Bulletin of Medical History* 4, no. 2 (1987); Gillian Rice, "The Bell-Magendie-Walker Controversy," *Medical History* 31, no. 2 (1987); and James Bradley, "Matters of Priority: Herbert Mayo, Charles Bell and Discoveries in the Nervous System," *Medical History* 58, no. 4 (2014).

50. For accounts of this encounter that favor Magendie, see Cranefield and Bell, *The Way In and the Way Out*; and Olmsted, *François Magendie, Pioneer in Experimental Physiology*. And for accounts that favor Bell, see Alexander Shaw, *An Account of Discoveries of Sir Charles Bell in the Nervous System* (London: J. Murray, 1860); and Pichot, *Life and Labours of Sir Charles Bell*.

51. Gordon-Taylor and Walls, *Sir Charles Bell, His Life and Times*.

52. Magendie, "Expériences sur les fonctions des racines des nerfs rachidiens," 366–71.

53. Translated in Olmsted, *François Magendie, Pioneer in Experimental Physiology*, 102.

54. Shaw and Bell referred to the "Idea of a New Anatomy" as having been published in 1809, although evidence suggests that it was not actually printed until 1811.

55. A narrative account of these events can be found in Magendie, "Expériences sur les fonctions des racines des nerfs qui naissent de la moëlle épinière."

56. Ibid. Translated in "The Discoveries of Sir Charles Bell," *London Medical Gazette* 21 (1838): 737–38.

57. Csiszar, "Seriality and the Search for Order"; and G. N. Cantor and Sally Shuttleworth, *Science Serialized: Representations of the Sciences in Nineteenth-Century Periodicals* (Cambridge: MIT Press, 2004).

58. Bell, *Exposition of the Natural System*.

59. Ibid., 3–4.

60. Desmond, *Politics of Evolution*; French, *Antivivisection and Medical Science*; and Rupke, *Vivisection in Historical Perspective*.

61. "Hansard's Parliamentary Reports, February 24" in 1825 states that "there was a Frenchman by the name of Magendie, whom he [Mr. Martin] considered a disgrace to Society . . . Mr. Martin added that he held in his hands the written declarations of Mr. Abernethy, of Sir Everard Home (and of other distinguished medical men), all uniting in condemnation of such excessive and protracted cruelty as had been practised by this Frenchman." See Hansard, "Hansard's Parliamentary Reports." French, *Antivivisection and Medical Science*; Rupke, *Vivisection in Historical Perspective*; and Guerrini, *Experimenting with Humans and Animals*.

62. Quoted in Shevawn Lynam, *Humanity Dick: A Biography of Richard Martin, M.P., 1754–1834* (London: Hamilton, 1975).

63. The list of animals protected was amended in 1835 to include dogs, chickens and roosters, bears, and bulls.

64. Bell, *Letters*, 265 (August 5, 1819).

65. Bell also considered himself heir to the traditions of the Hunters, whose Great Windmill Street School he took over. Bell wrote to his brother, "My object is . . . to make the town ring with it, as the only new thing that has appeared in anatomy since the days of Hunter." Ibid., 118 (December 5, 1807).

66. Warner, *Against the Spirit of System*, chap. 5.

67. Rehbock, "Transcendental Anatomy."

68. Bell, *Letters*, 281 (December 8, 1823).

69. For more on medical students going from Britain to France in the period following the Napoleonic Wars, see Bonner, *Becoming a Physician.*

70. Bell, "Lectures on the Nervous System."

71. Charles Bell, "Letter to the Editor On the Nervous System," *London Medical Gazette* 1 (1829).

72. Ibid.

73. For another version, see Bell, "Clinical Lecture on Diseases of the Nerves," 699. Bell writes: "When M. Magendie performed the experiments upon the spinal nerves, I saw that he went a great deal too far—farther than he was entitled to go by his premises. I saw that he was stating what he could not state from experiments, because his experiments were the same as mine. I had made out part of the subject—viz. that which related to the functions of the posterior roots—by inference, and then confirmed the whole by the decisive experiments upon the fifth pair. He pretended to make the thing clear by experiments upon those nerves which I had puzzled at in vain, in order to make clear by the very same experiments."

74. Ibid.

75. House of Commons, *Select Committee on Medical Education, Report*, 3 vols. (London: House of Commons, 1834), 2:216.

76. Charles Babbage, *Reflections on the Decline of Science in England* (London: B. Fellowes, 1830).

77. French, *Antivivisection and Medical Science*; and Roy Porter, *The Greatest Benefit to Mankind: A Medical History of Humanity* (New York: W. W. Norton, 1998), 335.

78. Bell, "Clinical Lecture on Diseases of the Nerves," 699.

79. "Higher anatomy" seems, for Bell, to have been a sort of anatomically based physiology—a physiology in which form followed function and in which there was a central and unified plan of nature. Philip Rehbock characterizes higher anatomy as being synonymous with philosophical anatomy and transcendental anatomy, forms of Romanticism that, Rehbock argues, made their way to Britain, in the form of eclectic and sometimes even contradictory philosophies, from Germany. Rehbock talks about Richard Owen as the prototypical transcendental anatomist, saying that he would have been "'The British Geoffroy' had it not been for the fact that he had already become known as 'the British Cuvier.' His approach was an eclectic one, employing transcendentalism and teleology as the situation warranted." Rehbock, "Transcendental Anatomy," 153.

80. Bell, *Letters*, 265 (August 5, 1819). See also Carin Berkowitz, "Establishing Authorship: Illustrative Style in Late Enlightenment Anatomy Folios," contribution to the special issue of *Bulletin of the History of Medicine: Anatomy, Aesthetics, and Authority* (forthcoming).

81. For more on the politics of medical journals during this period, see Desmond, *Politics of Evolution*, chaps. 3 and 4.

82. For more on the hierarchical system of medical professions and on the requirements for licentiates and fellows of the Colleges, see Lawrence, "Private Enterprise and Public Interests"; Roger French and Andrew Wear, *British Medicine in an Age of Reform*, Wellcome Institute Series in the History of Medicine (London: Routledge, 1991); Hamilton, "Medical Professions in the 18th Century"; Roy Porter, *Disease, Medicine, and Society in England, 1550–1860* (Basingstoke: Macmillan Education, 1987); and Christopher Lawrence, *Medicine in the Making of Modern Britain, 1700–1920* (London: Routledge, 1994).

83. Gordon-Taylor and Walls, *Sir Charles Bell, His Life and Times*; and Charles Herbert Mayo, *A Genealogical Account of the Mayo and Elton Families of the Counties of Wilts and Hereford* (London: C. Wittingham, 1882).

84. For more on the founding of the Middlesex Hospital School and on the overlap in Bell's and Mayo's careers, see Gordon-Taylor and Walls, *Sir Charles Bell, His Life and Times*, 146–49; and Wilson, *History of the Middlesex Hospital*, 165–69.

85. See, e.g., Charles Bell, "Observations on Fractures of the Patella," *London Medical Gazette* 1 (1827), and "Observations on the Diseases and Accidents to which the Hip Joint is Liable," *Lon-*

don Medical Gazette 1 (1828); and Herbert Mayo, "Dilated Oesophagus," *London Medical Gazette* 3 (1828).

86. Herbert Mayo, *Outlines of Human Physiology* (London: Burgess and Hill, 1827).

87. Herbert Mayo, "On Bellingeri's Claims as a Physiologist," *London Medical Gazette* (1834).

88. Mayo, *A Genealogical Account of the Mayo and Elton Families.*

89. Correspondent, "Medical Intelligence," *Provincial Medical and Surgical Journal* 19 (1852): 465.

90. Editor, "The Medical Session, School Arrangement," *London Medical Gazette* 18 (1836): 742.

91. Anonymous, "Obituary," *Lancet* 1852, vol. 2 (1852).

92. Mazumdar, "Anatomy, Physiology and Surgery," 141.

93. Shaw, *Narrative of the Discoveries of Sir Charles Bell.*

94. Charles Bell, "On the Nerves; Giving an Account of Some Experiments on Their Structure and Functions, Which Lead to a New Arrangement of the System," *Philosophical Transactions of the Royal Society of London* 111 (1821).

95. Charles Bell, *The Nervous System of the Human Body: Embracing the Papers Delivered to the Royal Society on the Subject of the Nerves* (London: Longman, Rees, Orme, Brown, and Green, 1830), 43.

96. Herbert Mayo and Johann Christian Reil, *Anatomical and Physiological Commentaries*, 2 vols. (London: printed for Thomas and George Underwood, 1822), 107–20.

97. Caesar Henry Hawkins, *The Hunterian Oration, Presidential Addresses, and Pathological and Surgical Writings*, 2 vols. (London: W. J. & S. Golbourn, 1874), 1:28.

98. For more on the politics of the various medical journals in the first half of the nineteenth century, see Desmond, *Politics of Evolution*, 13–16.

99. Herbert Mayo, "On the Uses of the Facial Branches of the 5th and 7th Nerves," *London Medical Gazette* 3 (1829): 831.

100. These were nerves controlling respiration—an essential function that required the coordination of many organs and muscles—which Bell believed occupied a distinct column called the "respiratory tract" in the spinal marrow and medulla oblongata.

101. Mayo, "On the Uses of the Facial Branches of the 5th and 7th Nerves."

102. Alexander Shaw, "Letter to the Editor, Mr. Shaw in Reply to Mr. Mayo," *London Medical Gazette* 4 (1829): 13.

103. Ibid.

104. Steven Shapin, "Of Gods and Kings: Natural Philosophy and Politics in the Leibniz-Clarke Disputes," *Isis* 72, no. 2 (1981); and Peter J. Bowler, *Charles Darwin: The Man and His Influence*, Cambridge Science Biographies Series (Cambridge: Cambridge University Press, 1996), 148–49.

105. Editor, "Controversy Concerning the Nervous System," *London Medical Gazette* 4 (1829): 60.

106. Bell, *Nervous System of the Human Body*, 188.

107. Anonymous Reviewer, "Review of Bell's *The Nervous System of the Human Body*," *London Medical Gazette* 7 (1831): 435.

108. "These things are contrived by cleverness, not by the power of knowledge." Herbert Mayo, "To the Editor of the *Medical Quarterly Review*," *Medical Quarterly Review* 2 (1834): 451.

109. Shaw, *Narrative of the Discoveries of Sir Charles Bell.*

110. Ibid., 45.

111. Ibid., 7.

112. Bell, "Mr. Bell's Letter to his Pupils of the London University, on Taking Leave of Them."

113. Shaw, *Narrative of the Discoveries of Sir Charles Bell*, 7.

114. For more on the birth of physiology as an experimental discipline, see Andrew Cunningham, "The Pen and the Sword: Recovering the Disciplinary Identity of Physiology and Anatomy

Before 1800. I, Old Physiology—the Pen," *Studies in History and Philosophy of Biological and Biomedical Sciences* 33C, no. 4 (2002), and "The Pen and the Sword: Recovering the Disciplinary Identity of Physiology and Anatomy before 1800: II, Old Anatomy—the Sword," *Studies in History and Philosophy of Biological and Biomedical Sciences* 34C, no. 1 (2003).

115. Shaw, *Narrative of the Discoveries of Sir Charles Bell*, 7.

116. Warner, "Idea of Science in English Medicine," 138.

117. Rupke, *Vivisection in Historical Perspective*, 92–102; and French, *Antivivisection and Medical Science*, 15–35.

118. Mazumdar, "Anatomy, Physiology and Surgery"; and Porter, "Medical Lecturing in Georgian London."

119. For a parallel story, see Iwan Rhys Morus, *Frankenstein's Children: Electricity, Exhibition, and Experiment in Early-Nineteenth-Century London* (Princeton: Princeton University Press, 1998). Electrical science attracted overlapping but distinct communities of spectacle seekers, technicians, and scientists, like Michael Faraday at the Royal Institution.

120. Bell's palsy is named after Charles Bell, who used patients' partial paralysis and his experiences cutting facial nerves to alleviate their pain to help illuminate the normal motor and sensory functions of facial nerves. Bell, *Nervous System of the Human Body*.

EPILOGUE

1. Bell, *Nervous System of the Human Body*, preface, xxi.

2. Steven Shapin and Adi Ophir, "The Place of Knowledge: A Methodological Survey," *Science in Context* 4, no. 1 (1991).

3. Bell, *Letters*, 410.

4. Ibid., 316 (December 8, 1835).

5. Mayo and Reil, *Anatomical and Physiological Commentaries*, 112.

6. Ibid., 114.

7. "By the experiments and reasoning which I have described, I thus established that the ganglionless portion of the fifth and the hard portion of the seventh nerve are voluntary nerves to the parts, which receive sentient nerves from the larger or ganglionic portion of the fifth. This happened before the publication of M. Magendie's discovery of the parallel functions of the double roots of the spinal nerves; and without wishing to assert the least claim to that discovery, I will yet observe, that I was led by the well-known anatomical analogy between the fifth and spinal nerves, to conjecture nearly what M. Magendie proved, and was indeed actually engaged in experiments to determine the point, when M. Magendie's were published." Herbert Mayo, *Outlines of Human Physiology*, 3rd ed. (London: Burgess and Hill, 1833), 262.

8. Shaw, *Narrative of the Discoveries of Sir Charles Bell*, 76.

9. Ibid., 23.

10. Mayo, *Outlines of Human Physiology*, 6–7.

11. Ibid., iii.

12. Ibid., iv.

13. Geison, *Michael Foster and the Cambridge School of Physiology*.

14. Quoted in Geison, *Michael Foster and the Cambridge School of Physiology*, 189.

15. Ibid., 186.

16. Ibid., 187.

17. Bell, *Letters*, 413.

18. Ibid., 252 (December 15, 1815).

Bibliography

Abernethy, John. *Lectures on the Theory and Practice of Surgery*. London: Longman, Rees, Orme, Brown, and Green, 1830.

Ackerknecht, Erwin. *Medicine at the Paris Hospital, 1794–1848*. Baltimore: Johns Hopkins University Press, 1967.

Åhrén, Eva. *Death, Modernity, and the Body: Sweden 1870–1940*. Rochester: University of Rochester Press, 2009.

Alberti, Samuel J. M. M. *Morbid Curiosities: Medical Museums in Nineteenth-Century Britain*. Oxford: Oxford University Press, 2011.

Alberti, Samuel J. M. M., and Elizabeth Hallam. *Medical Museums: Past, Present, Future*. London: Royal College of Surgeons of England, 2013.

Anonymous. "Anonymous Fair-Copy Notes from Lectures on Anatomy and Surgery Reputedly Delivered by Sir Charles Bell in 1822." Brotherton Special Collections, Leeds University Library.

———. "Charles Bell's Bridgewater Treatise." *London Medical Gazette* 13 (1833): 253–58.

———. "Letter to the Editor: Collegium Wakleyanum." *London Medical Gazette* 8 (1831): 271–72.

———. "Letter to the Editor: London College of Medicine." *Lancet* 1830–31, vol. 2 (1831): 256.

———. "Letters and Discoveries of Sir Charles Bell." *Edinburgh Review* 136, no. 1872 (1872): 394–429.

———. "London College of Medicine, Second Public Meeting of the Supporters of This Institution." *Lancet* 1830–31, vol. 2 (1831): 212–22.

———. "Museum Making." *Lancet* 1826–27, vol. 2 (1827): 374.

———. "Obituary." *Lancet* 1852, vol. 2 (1852): 313.

———. "Review of Bell's *The Nervous System of the Human Body*." *London Medical Gazette* 7 (1831): 434–37.

Anonymous ("Argus"). "Letter to the Editor: Medical and Surgical Reform." *Lancet* 1828–29, vol. 2 (1828): 397.

Anonymous ("Juvenis"). "Apothecaries." *Lancet* 1828–29, vol. 2 (1828): 429.

Anonymous ("L."). "London College of Medicine: Objections to the Universal Title of 'Doctor.'" *Lancet* 1830–31, vol. 2 (1831): 687–88.

Anonymous ("Omega"). "Letter to the Editor: London College of Medicine." *Lancet* 1830–31, vol. 2 (1831): 345–46.

Anonymous ("One of the Multitude"). "College of Surgeons." *London Medical Gazette* 7 (1831): 818–20.

Anonymous ("Philomeides"). "Letter to the Editor: Grand Convocation of the Collegium Wakleyanum." *London Medical Gazette* 8 (1831): 208–9.

Anonymous ("Voyageur"). "Letter to the Editor: Parisian Medicine." *London Medical Gazette* 1 (1828): 695–97.

Appel, Toby A. *The Cuvier-Geoffroy Debate: French Biology in the Decades before Darwin*. New York: Oxford University Press, 1987.

Babbage, Charles. *Reflections on the Decline of Science in England*. London: B. Fellowes, 1830.

Baillie, Matthew. *The Morbid Anatomy of Some of the Most Important Parts of the Human Body*. London: J. Johnson, 1793.

———. *A Series of Engravings, Accompanied with Explanations, Which Are Intended to Illustrate the Morbid Anatomy of Some of the Most Important Parts of the Human Body*. London: printed by W. Bulmer and Co. for J. Johnson; and G. and W. Nicol, 1799.

Barfoot, Michael. "Philosophy and Method in Cullen's Medical Teaching." In *William Cullen and the Eighteenth-Century Medical World*, edited by A. Doig, J. P. S. Ferguson, I. A. Milne, and R. Passmore, 110–32. Edinburgh: Edinburgh University Press, 1993.

Bell, Sir Charles. "Clinical Lecture on Diseases of the Nerves." *London Medical Gazette* 13 (1833–34): 699.

———. "Clinical Lecture on Diseases of the Spine, Delivered, Nov. 3, at the Middlesex Hospital School." *London Medical Gazette* 17 (1835): 231–34.

———. *Engravings of the Arteries, Illustrating the Second Volume of the Anatomy of the Human Body*, edited by John Bell. London: Longman and Rees, 1801.

———. *Essays on the Anatomy of Expression in Painting*. London: Longman, Hurst, Rees, and Orme, 1806.

———. *Essays on the Anatomy and Philosophy of Expression*. London: John Murray, 1824.

———. *An Exposition of the Natural System of the Nerves of the Human Body*. London: Spottiswode, 1825.

———. *The Hand, Its Mechanism and Vital Endowments as Evincing Design*. London: William Pickering, 1833.

———. *Idea of a New Anatomy of the Brain*. London, 1811.

———. "Idea of a New Anatomy of the Brain, with Letters &C.." *Journal of Anatomy and Physiology* 3, no. 1 (1868): 147–82.

———. *Idea of a New Anatomy of the Brain: A Facsimile of the Privately Printed Edition of* 1811, *with a Bio-Bibliographical Introduction*. London: Dawsons of Pall Mall, 1966.

———. *Institutes of Surgery; Arranged in the Order of the Lectures Delivered in the University of Edinburgh*. 2 vols. Edinburgh: Black, 1838.

———. "Lectures on the Nervous System Delivered at the College of Surgeons." *London Medical Gazette* 1 (1828): 617–22.

———. "Letter to the Editor on the Nervous System." *London Medical Gazette* 3 (1829): 691–92.

———. *Letters Concerning the Diseases of the Urethra*. London: Longman, 1810.

———. *Letters of Sir Charles Bell, Selected from His Correspondence with His Brother George Joseph Bell*. London: J. Murray, 1870.

———. "London University—Mr. Bell's Introductory Lecture." *London Medical Gazette* 5 (1830): 18–21.

———. "Middlesex Hospital School: Opening Address." *Lancet* 1835–36, vol. 1 (1835): 89.

———. "Mr. Bell's Letter to His Pupils of the London University, on Taking Leave of Them." *London Medical Gazette* 7 (1830): 308–11.

———. *The Nervous System of the Human Body: Embracing the Papers Delivered to the Royal Society on the Subject of the Nerves*. London: Longman, Rees, Orme, Brown, and Green, 1830.

———. "Observations on Fractures of the Patella." *London Medical Gazette* 1 (1827): 28–31.

———. "Observations on the Diseases and Accidents to Which the Hip Joint Is Liable." *London Medical Gazette* 1 (1828): 137–42.

———. "On the Nerves; Giving an Account of Some Experiments on Their Structure and Functions, Which Lead to a New Arrangement of the System." *Philosophical Transactions of the Royal Society of London* 111 (1821): 398–424.

———. "Sir Charles Bell's Introductory Address on the Opening of London University." *London Medical Gazette* 2 (1828): 566–68.

———. *A System of Operative Surgery: Founded on the Basis of Anatomy*. 2 vols. London: Longman, Hurst, Rees, and Orme, 1807.

Bell, John. *Engravings of the Bones, Muscles, and Joints*. London: Longman and Rees, and Cadell and Davies, 1804.

———. *Letters on Professional Character and Manners: On the Education of a Surgeon, and the Duties and Qualifications of a Physician: Addressed to James Gregory, M.D.* Edinburgh: John Moir, 1810.

Bell, John, and Charles Bell. *The Anatomy of the Human Body*. 4 vols. London: T. N. Longman and O. Rees [etc.], 1802.

Berkowitz, Carin. "The Beauty of Anatomy: Visual Displays and Surgical Education in Early Nineteenth-Century London." *Bulletin of the History of Medicine* 85, no. 2 (2011): 248–71.

———. "Disputed Discovery: Vivisection and Experiment in the 19th Century." *Endeavour* 30, no. 3 (2006): 98–102.

———. "Defining a Discovery: Priority and Methodological Controversy in Early Nineteenth-Century Anatomy." *Notes and Records of the Royal Society* 68, no. 4 (2014): in press.

———. "Establishing Authorship: Illustrative Style in Late Enlightenment Anatomy Folios." Contribution to special issue of *Bulletin of the History of Medicine: Anatomy, Aesthetics, and Authority* (forthcoming).

———. "Medical Science as Pedagogy in Early Nineteenth-Century Britain: Charles Bell and the Politics of London Medical Reform." PhD dissertation, Cornell University, 2010.

———. "Systems of Display: The Making of Anatomical Knowledge in Enlightenment Britain." *British Journal for the History of Science* 46, no. 3 (2013): 359–87.

Bevan, Michael. "Cline, Henry (1750–1827)." *Oxford Dictionary of National Biography*. Oxford: Oxford University Press (2004). http://www.oxforddnb.com/view/article/5673.

Bonner, Thomas Neville. *Becoming a Physician: Medical Education in Britain, France, Germany, and the United States, 1750–1945*. Oxford: Oxford University Press, 1995.

Bostetter, Mary. "The Journalism of Thomas Wakley." In *Innovators and Preachers: The Role of the Editor in Victorian England*, edited by Joel H. Wiener, 265–92. Westport: Greenwood, 1985.

Bowler, Peter J. *Charles Darwin: The Man and His Influence*. Cambridge Science Biographies Series. Cambridge: Cambridge University Press, 1996.

Bradley, James. "Matters of Priority: Herbert Mayo, Charles Bell and Discoveries in the Nervous System." *Medical History* 58, no. 4 (2014): 564–84.

Brannigan, Augustine. *The Social Basis of Scientific Discoveries*. Cambridge: Cambridge University Press, 1981.

Broadie, Alexander, ed. *The Cambridge Companion to the Scottish Enlightenment*. Cambridge: Cambridge University Press, 2003.

Brock, R. C. *The Life and Work of Astley Cooper*. Edinburgh: E & S Livingstone Ltd., 1952.

Brockliss, L. W. B. "Before the Clinic: French Medical Teaching in the Eighteenth Century." In *Constructing Paris Medicine*, edited by Caroline Hannaway and Ann La Berge, 71–115. Amsterdam: Rodopi, 1998.

Brockliss, L. W. B., and Colin Jones. *The Medical World of Early Modern France*. Oxford: Oxford University Press, 1997.

Brodie, Benjamin. "Introductory Discourse on the Studies Required for the Medical Profession." In *The Works of Sir Benjamin Collins Brodie*, edited by Charles Hawkins, 465–84. London: Longman, Green, Longman, Roberts, and Green, 1838.

———. *Pathological and Surgical Observations on Diseases of the Joints*. London: Longman, Hurst, Rees, Orme, and Brown, 1818.

Broman, Thomas. "The Habermasian Public Sphere and 'Science in the Enlightenment.'" *History of Science* 36 (1998): 123–49.

———. "J. C. Reil and the 'Journalization' of Physiology." In *The Literary Structure of Scientific Argument: Historical Studies*, edited by Peter Dear, vi, 211. Philadelphia: University of Pennsylvania Press, 1991.

Brown, Michael. "'Bats, Rats and Barristers': *The Lancet*, Libel and the Radical Stylistics of Early Nineteenth-Century English Medicine," *Social History* 39 (2014): 182–209.

Brown, Stuart C. *British Philosophy and the Age of Enlightenment*. London: Routledge, 1996.

Burke, Edmund. "The London University." *Annual Register* 67 (1826): 82.

Burney, Ian. "Medicine in the Age of Reform." In *Rethinking the Age of Reform: Britain 1780–1850*, edited by Arthur Burns and Joanna Innes, 163–81. Cambridge: Cambridge University Press, 2003.

Burns, Arthur, and Joanna Innes. *Rethinking the Age of Reform: Britain 1780–1850*. Cambridge; New York: Cambridge University Press, 2003.

Buttmann, Günther, and David S. Evans. *The Shadow of the Telescope: A Biography of John Herschel*. Guildford: Lutterworth Press, 1974.

Bynum, W. F. *Science and the Practice of Medicine in the Nineteenth Century*. Cambridge: Cambridge University Press, 1994.

Bynum, W. F., and Roy Porter. *William Hunter and the Eighteenth-Century Medical World*. Cambridge: Cambridge University Press, 1985.

Bynum, William, and Janice Wilson. "Periodical Knowledge: Medical Journals and Their Editors in Nineteenth-Century Britain." In *Medical Journals and Medical Knowledge: Historical Essays*, edited by William Bynum, Stephen Lock, and Roy Porter, chap. 2. London: Routledge, 1992.

Cantor, G. N., and Sally Shuttleworth. *Science Serialized: Representations of the Sciences in Nineteenth-Century Periodicals*. Cambridge: MIT Press, 2004.

Cantor, Geoffrey. "Henry Brougham and the Scottish Methodological Tradition." *Studies in History and Philosophy of Science* 2 (1971): 69–89.

———. *Science in the Nineteenth-Century Periodical: Reading the Magazine of Nature*. Cambridge Studies in Nineteenth-Century Literature and Culture. Cambridge: Cambridge University Press, 2004.

Chaplin, Simon. "Nature Dissected, or Dissection Naturalized? The Case of John Hunter's Museum." *Museum & Society* 6 (2008): 135–51.

Cheselden, W. "An Account of some Observations made by a young Gentleman, who was born blind, or lost his Sight so early, that he had no Remembrance of ever having seen, and was couch'd between 13 and 14 Years of Age." *Philosophical Transactions* 402 (1728): 447–50.

Christie, J. R. R. "Adam Smith's Metaphysics of Language." In *The Figural and the Literal: Problems of Language in the History of Science and Philosophy*, 1630–1800, edited by Geoffrey N. Cantor, Andrew E. Benjamin, and J. R. R. Christie, 202–29. Manchester: Manchester University Press, 1987.

———. "Ether and the Science of Chemistry: 1740–1790." In *Conceptions of Ether: Studies in the History of Ether Theories*, 1740–1900, edited by Jonathan Hodge and Geoffrey Cantor, 85–111. Cambridge: Cambridge University Press, 1981.

———. "The Origins and Development of the Scottish Scientific Community, 1680–1760." *History of Science* 12 (1974): 122–41.

Clarke, Edwin, and L. S. Jacyna. *Nineteenth-Century Origins of Neuroscientific Concepts*. Berkeley: University of California Press, 1987.

Cohen, John. "Medical Education in the University of London, University College and Middlesex Hospitals 1800–1840." MA thesis, University of London, 1991.

Coleman, William, and Frederic Lawrence Holmes. *The Investigative Enterprise: Experimental Physiology in Nineteenth-Century Medicine*. Berkeley: University of California Press, 1988.

Colley, Linda. *Britons: Forging the Nation, 1707–1837*. New Haven: Yale University Press, 1992.

Collier, Bruce, and James H. MacLachlan. *Charles Babbage and the Engines of Perfection*. Oxford Portraits in Science. New York: Oxford University Press, 1998.

Cooper, Astley. *The Lectures of Sir Astley Cooper, Bart. On the Principles and Practice of Surgery*. 3 vols. Boston: Wells and Lilly, 1825.

Cooper, Bransby Blake. *The Life of Sir Astley Cooper, Bart., Interspersed with Sketches from His Note-Books of Distinguished Contemporary Characters*. London: J. W. Parker, 1843.

Cooter, Roger. *The Cultural Meaning of Popular Science: Phrenology and the Organization of Consent in Nineteenth-Century Britain*. Cambridge History of Medicine. Cambridge: Cambridge University Press, 1984.

———. *Phrenology in the British Isles: An Annotated Historical Biobibliography and Index*. Metuchen: Scarecrow Press, 1989.

Correspondent. "Medical Intelligence." *Provincial Medical and Surgical Journal* 19 (1852): 464–5.

Cranefield, Paul F., and Charles Bell. *The Way In and the Way Out: François Magendie, Charles Bell, and the Roots of the Spinal Nerves: With a Facsimile of Charles Bell's Annotated Copy of His Ideas of a New Anatomy of the Brain*. Mount Kisco: Futura Publishing, 1974.

Csiszar, Alex. "Seriality and the Search for Order: Scientific Print and Its Problems During the Late Nineteenth Century." *History of Science* 48, no. 3/4 (2010): 399–434.

Cummings, Frederick. "B. R. Haydon and His School." *Journal of the Warburg and Courtauld Institutes* 26, no. 3/4 (1963): 367–80.

Cunningham, Andrew. *The Anatomist Anatomis'd: An Experimental Discipline in Enlightenment Europe*. Farnham; Burlington: Ashgate, 2010.

———. "The Pen and the Sword: Recovering the Disciplinary Identity of Physiology and Anatomy before 1800. I, Old Physiology—the Pen." *Studies in History and Philosophy of Biological and Biomedical Sciences* 33C (2002): 631–65.

———. "The Pen and the Sword: Recovering the Disciplinary Identity of Physiology and Anatomy before 1800: II, Old Anatomy—the Sword." *Studies in History and Philosophy of Biological and Biomedical Sciences* 34C (2003): 51–76.

Cunningham, Andrew, and Perry Williams. *The Laboratory Revolution in Medicine*. Cambridge: Cambridge University Press, 1992.

Daston, Lorraine, and Peter Galison. "The Image of Objectivity." *Representations* 40 (1992): 81–128.

———. *Objectivity*. Cambridge: MIT Press, 2007.

Dear, Peter. *The Intelligibility of Nature: How Science Makes Sense of the World*. Chicago: University of Chicago Press, 2006.

Desmond, Adrian J. *The Politics of Evolution: Morphology, Medicine, and Reform in Radical London*. Chicago: University of Chicago Press, 1989.

"The Discoveries of Sir Charles Bell." *London Medical Gazette* 21 (1838): 734–38.

Donkin, H. B., and C. MacNamara. *The Westminster Hospital Reports*. Vol. 1. London: J. E. Adlard, 1885.

Donovan, Arthur. *Philosophical Chemistry in the Scottish Enlightenment*. Edinburgh: Edinburgh University Press, 1984.

Eddy, Matthew. "Converging Paths or Separate Roads? The Roles Played by Science, Medicine, and Philosophy in the Scottish Enlightenment." *Philosophical Writings* 30 (2005): 30–40.

———. "Nineteenth-Century Natural Theology." In *Oxford Handbook of Natural Theology*, edited by Russell Manning, 101–17. Oxford: Oxford University Press, 2013.

———. *The Patchwork Picture: Science, Education and the Visual Foundations of Knowledge, 1760–1820*. Chicago: University of Chicago Press, forthcoming.

"Editorial: London University." *London Magazine* 5 (August, 1826): 554.

"Editorial: London University—Mr. Bell." *London Medical Gazette* 7 (1830): 308–11.

"Editorial: Resignation of Mr. C. Bell." *London Medical Gazette* 6 (1830): 469–73.

Elliotson, John. "Introductory Address to a Course of Lectures on the Principles and Practice of Medicine." *Lancet* 1831–32, vol. 1 (1831): 64–70.

Evans, Eric J. *Britain before the Reform Act: Politics and Society 1815–1832*. Seminar Studies in History. London: Longman, 1989.

———. *The Forging of the Modern State: Early Industrial Britain, 1783–1870*. Foundations of Modern Britain. London: Longman, 1983.

———. *The Great Reform Act of 1832*. Lancaster Pamphlets. London: Methuen, 1983.

Flourens, Pierre. "1858 Memoir of Magendie, Translated by C. A. Alexander." *Annual Report of the Smithsonian Institution* (1866): 100.

Foucault, Michel. *The Birth of the Clinic: An Archaeology of Medical Perception*. New York: Pantheon Books, 1973.

French, Richard D. *Antivivisection and Medical Science in Victorian Society*. Princeton: Princeton University Press, 1975.

French, Roger. *William Harvey's Natural Philosophy*. Cambridge: Cambridge University Press, 1994.

French, Roger, and Andrew Wear. *British Medicine in an Age of Reform*. Wellcome Institute Series in the History of Medicine. London: Routledge, 1991.

Fry, Michael. "Jeffrey, Francis, Lord Jeffrey (1773–1850)." In *Oxford Dictionary of National Biography*. Oxford: Oxford University Press, 2004. http://www.oxforddnb.com/index/101014698/Francis-Jeffrey.

Fyfe, Aileen. "Publishing and the Classics: Paley's Natural Theology and the Nineteenth-Century Scientific Canon." *Studies in History and Philosophy of Science* 33 (2002): 729–51.

———. *Science and Salvation: Evangelical Popular Science Publishing in Victorian Britain*. Chicago: University of Chicago Press, 2004.

Gascoigne, John. *Joseph Banks and the English Enlightenment: Useful Knowledge and Polite Culture*. Cambridge: Cambridge University Press, 1994.

———. *Science in the Service of Empire: Joseph Banks, the British State and the Uses of Science in the Age of Revolution*. Cambridge: Cambridge University Press, 1998.

Geison, Gerald. "Divided We Stand: Physiologists and Clinicians in the American Context." In *The Therapeutic Revolution: Essays in the Social History of American Medicine*, edited by Morris J. Vogel and Charles Rosenberg, 67–90. Philadelphia: University of Pennsylvania Press, 1979.

———. *Michael Foster and the Cambridge School of Physiology: The Scientific Enterprise in Late Victorian Society*. Princeton: Princeton University Press, 1978.

Geison, Gerald, and Frederic Holmes, eds. *Research Schools: Historical Reappraisals*. *Osiris* 8. Chicago: University of Chicago Press, 1993.

Gelfand, Toby. *Professionalizing Modern Medicine: Paris Surgeons and Medical Science and Institutions in the 18th Century*. Westport: Greenwood Press, 1980.

Golinski, Jan. "Humphry Davy: The Experimental Self." *Eighteenth-Century Studies* 45, no. 1 (2010): 15–28.

———. *Making Natural Knowledge: Constructivism and the History of Science*. Cambridge History of Science. Cambridge: Cambridge University Press, 1998.

Gordin, Michael D. *A Well-Ordered Thing: Dmitrii Mendeleev and the Shadow of the Periodic Table*. New York: Basic Books, 2004.

Gordon-Taylor, Gordon, and E. W. Walls. *Sir Charles Bell, His Life and Times*. Edinburgh: E. & S. Livingstone, 1958.

Green, John Alfred. *The Educational Ideas of Pestalozzi*. New York: Greenwood Press, 1969.

Green, Joseph Henry. "Distinction without Separation." *London Medical Gazette* 8 (1831): 213–17.

Guerrini, Anita. "The Ethics of Animal Experimentation in Seventeenth-Century England." *Journal of the History of Ideas* 50 (1989): 391–407.

———. *Experimenting with Humans and Animals: From Galen to Animal Rights*. Johns Hopkins Introductory Studies in the History of Science. Baltimore: Johns Hopkins University Press, 2003.

H. R. "Some Account of Pestalozzi and His Method of Instruction." *Athenaeum* 2, no. 10 (1807).

Hamilton, Bernice. "The Medical Professions in the 18th Century." *Economic History Review* 4, no. 2 (1951): 141–69.

Hansard, T. C. "Hansard's Parliamentary Reports, February 24." London, 1825.

Harris, James A. *Of Liberty and Necessity: The Free Will Debate in Eighteenth-Century British Philosophy*. Oxford Philosophical Monographs. Oxford: Oxford University Press, 2005.

Harte, N. B. *The University of London, 1836–1986: An Illustrated History*. London: Athlone Press, 1986.

Harwood, Jonathan. *Styles of Scientific Thought: The German Genetics Community, 1900–1933*. Science and Its Conceptual Foundations. Chicago: University of Chicago Press, 1993.

Hawkins, Caesar Henry. *The Hunterian Oration, Presidential Addresses, and Pathological and Surgical Writings*. 2 vols. London: W. J. & S. Golbourn, 1874.

Herschel, John F. W., and David S. Evans. *Herschel at the Cape: Diaries and Correspondence of Sir John Herschel, 1834–1838*. History of Science Series 1. Austin: University of Texas Press, 1969.

Hobsbawm, E. J., and T. O. Ranger. *The Invention of Tradition*. Cambridge; New York: Cambridge University Press, 1983.

Holloway, S. W. F. "The Apothecaries Act." *Medical History* 10, no. 3 (1966): 221–36.

Horner, Leonard. "University of London, Medical Diploma." *London Medical Gazette* 6 (1830): 219–21.

House of Commons, *Select Committee on Medical Education, Report*, 3 vols. (London: House of Commons, 1834), 2:216.

Hunt, Richard, and Ida Macalpine. "A Privately Printed Pamphlet by Sir Charles Bell on the Principles Involved in Appointments to the London Hospitals." *Annals of the Royal College of Surgeons of England* 30 (1962): 257–65.

Hunter, William. *Anatomia uteri humani gravidi tabulis illustrata*. Birmingham: John Baskerville, 1774.

——. *Two Introductory Lectures, Delivered by Dr. William Hunter to His Last Course of Anatomical Lectures, at His Theatre in Great Windmill Street*. London: J. Johnson, 1784.

Hunter, William, Alice Julia Marshall, and John H. Teacher. *Catalogue of the Anatomical Preparations of William Hunter in the Museum of the Anatomy Department*. Glasgow: University of Glasgow, 1970.

Hunting, Penelope. *History of the Royal Society of Medicine*. London: Royal Society of Medicine, 2001.

Hyman, Anthony. *Charles Babbage, Pioneer of the Computer*. Oxford: Oxford University Press, 1984.

Jacyna, L. S. "Bell, Sir Charles (1774–1842)." In *Oxford Dictionary of National Biography*. Oxford: Oxford University Press, 2004. http://www.oxforddnb.com/index/101001999/Charles-Bell.

——. "Images of John Hunter in the Nineteenth Century." *History of Science* 21 (1983): 85—108.

——. "Immanence or Transcendence: Theories of Life and Organization in Britain 1790–1835." *Isis* 74 (1983): 311–29.

——. *Philosophic Whigs: Medicine, Science, and Citizenship in Edinburgh, 1789–1848*. London: Routledge, 1994.

——. "Theory of Medicine, Science of Life: The Place of Physiology Teaching in the Edinburgh Medical Curriculum, 1790–1870." In *The History of Medical Education in Britain*, edited by Vivian Nutton and Roy Porter, 141–53. Amsterdam: Rodopi BV Editions, 1995.

Jewson, Nicholas. "Medical Knowledge and the Patronage System in 18th Century England." *Sociology* 8, no. 3 (1974): 369–85.

"John Bell." In *Penny Cyclopedia of the Society for the Diffusion of Useful Knowledge*. London: Charles Knight, 1835.

Jordanova, L. J. *The Look of the Past: Visual and Material Evidence in Historical Practice*. Cambridge: Cambridge University Press, 2012.

Jordanova, Ludmilla. "Gender, Generation and Science: William Hunter's Obstetrical Atlas." In

William Hunter and the Eighteenth Century Medical World, edited by William Bynum and Roy Porter, 385–417. Cambridge: Cambridge University Press, 1985.

———. "Medicine and Genres of Display." In *Visual Display: Culture Beyond Appearances*, edited by L. Cooke and P. Wollen, 202–17. Seattle: Bay Press, 1995.

———. "The Representation of the Human Body: Art and Medicine in the Work of Charles Bell." In *Towards a Modern Art World*, edited by Brian Allen, 79–94. New Haven: Yale University Press, 1995.

Jorgensen, C. B. "Aspects of the History of the Nerves: Bell's Theory, the Bell-Magendie Law and Controversy, and Two Forgotten Works by P. W. Lund and D. F. Eschricht." *Journal of the History of Neuroscience* 12, no. 3 (2003): 229–49.

Kaufman, Matthew. *Robert Liston: Surgery's Hero*. Edinburgh: Royal College of Surgeons of Edinburgh, 2009.

Keene, Melanie. "Object Lessons: Sensory Science Education, 1830–1870." PhD thesis, University of Cambridge, 2009.

Kemp, Martin. *Dr. William Hunter at the Royal Academy of Arts*. Glasgow: University of Glasgow Press, 1975.

———. "'The Mark of Truth': Looking and Learning in Some Anatomical Illustrations from the Renaissance and the Eighteenth Century." In *Medicine and the Five Senses*, edited by William Bynum and Roy Porter, 85–121. Cambridge: Cambridge University Press, 1993.

———. "Style and Non-Style in Anatomical Illustration: From Renaissance Humanism to Henry Gray." *Journal of Anatomy* 216, no. 2 (2010): 192–208.

———. "True to Their Natures: Sir Joshua Reynolds and Dr. William Hunter at the Royal Academy of Arts." *Notes and Records of the Royal Society of London* 46 (1992): 77–88.

Kemp, Martin, and Marina Wallace. *Spectacular Bodies: The Art and Science of the Human Body from Leonardo to Now*. Berkeley: University of California Press, 2000.

Kidd, Colin. *British Identities before Nationalism: Ethnicity and Nationhood in the Atlantic World, 1600–1800*. Cambridge: Cambridge University Press, 1999.

Kinraide, Rebecca. "The Society for the Diffusion of Useful Knowledge and the Democratization of Learning in Early Nineteenth-Century Britain." PhD thesis, University of Wisconsin-Madison, 2006.

Kusukawa, Sachiko. *Picturing the Book of Nature: Image, Text, and Argument in Sixteenth-Century Human Anatomy and Medical Botany*. Chicago: University of Chicago Press, 2012.

Lancet. "Introductory Address on the Opening of the Middlesex Hospital Medical School." *Lancet* 1835–36, vol. 1 (1835): 148.

———. "St. Bartholomew's Hospital." *Lancet* 1829–30, vol. 1 (1829): 47.

Lawrence, Christopher. "Alexander Monro Primus and the Edinburgh Manner of Anatomy." *Bulletin of the History of Medicine* 62 (1988): 193–214.

———. "The Edinburgh Medical School and the End of the 'Old Thing,' 1790–1830." *History of Universities* 7 (1988): 259–86.

———. *Medical Theory, Surgical Practice: Studies in the History of Surgery*. London: Routledge, 1992.

———. *Medicine in the Making of Modern Britain, 1700–1920*. Historical Connections. London: Routledge, 1994.

———. "The Nervous System and Society in the Scottish Enlightenment." In *Natural Order: Historical Studies of Scientific Culture*, edited by B. Barnes and S. Shapin, 19–40. Beverly Hills: Sage Publications, 1979.

———. "The Shaping of Things to Come: Scottish Medical Education 1700–1939." *Medical Education* 40, no. 3 (2006): 212–18.

———. "Medicine as Culture: Edinburgh and the Scottish Enlightenment." PhD thesis, University of London, 1984.

Lawrence, Susan C. *Charitable Knowledge: Hospital Pupils and Practitioners in Eighteenth-Century London*. Cambridge: Cambridge University Press, 1996.

———. "Educating the Senses: Students, Teachers and Medical Rhetoric in Eighteenth-Century

London." In *Medicine and the Five Senses*, edited by W. F. Bynum and Roy Porter, 154–78. Cambridge: Cambridge University Press, 1993.

———. "Entrepreneurs and Private Enterprise: The Development of Medical Lecturing in London, 1775–1820." *Bulletin of the History of Medicine* 62 (1988): 171–92.

———. "Private Enterprise and Public Interests: Medical Education and the Apothecaries' Act, 1780–1825." In *British Medicine in an Age of Reform*, edited by Roger French and Andrew Wear, 45–74. London: Routledge, 1991.

Le Fanu, William."Sir Benjamin Brodie." *Notes and Records of the Royal Society of London* 19, no. 1 (1964): 42–52.

Lesch, John E. *Science and Medicine in France: The Emergence of Experimental Physiology, 1790–1855*. Cambridge: Harvard University Press, 1984.

Locke, John. *An Essay Concerning Human Understanding*. London: printed by Eliz. Holt for Thomas Bassett, 1690.

London Medical Gazette. "College of Surgeons." *London Medical Gazette* 7 (1831): 787–91.

———. "Controversy Concerning the Nervous System." *London Medical Gazette* 4 (1829): 60.

———. "Cooper v. Wakley." *London Medical Gazette* 3 (1828): 65–98.

———. "Criminal Information against the Rioters—New Bye-Laws of the College of Surgeons." *London Medical Gazette* 8 (1831): 279.

———. "Hospital Reporting." *London Medical Gazette* 1 (1828): 697–701.

———. "Hospital Reporting." *London Medical Gazette* 2 (1828): 120–21.

———. "London University—Apprenticeships." *London Medical Gazette* 8 (1831): 336.

———. "Medical Education." *London Medical Gazette* 1 (1828): 314–17.

———. "Medical Reform—Education." *London Medical Gazette* 11 (1832): 89–92.

———. "The Medical Session, School Arrangement." *London Medical Gazette* 18 (1836): 741–42.

———. "Memorial of the Medical Teachers." *London Medical Gazette* 14 (1834): 241–44.

———. "Present State of the London and Paris Schools of Medicine." *London Medical Gazette* 7 (1830): 21–25.

———. "Reform—College of Physicians." *London Medical Gazette* 11 (1832–33): 485.

London Medical and Surgical Journal. "Medical Journalism—Notice of Parliamentary Inquiry." *London Medical and Surgical Journal* 5 (1834): 55–57.

Loudon, Irvine. *Medical Care and the General Practitioner, 1750–1850*. Oxford: Clarendon Press, 1986.

———. "Medical Practitioners 1750–1850 and the Period of Medical Reform in Britain." In *Medicine in Society*, edited by Andrew Wear, 219–48. Cambridge: Cambridge University Press, 1992.

———. "The Nature of Provincial Medical Practice in Eighteenth-Century England." *Medical History* 29 (1985): 1–32.

Loudon, Irving. "Sir Charles Bell and the Anatomy of Expression." *British Medical Journal* 285 (1982): 1794–96.

Lynam, Shevawn. *Humanity Dick: A Biography of Richard Martin, M.P., 1754–1834*. London: Hamilton, 1975.

Macaulay, Thomas Babington. "Thoughts on the Advancement of Academical Education in England." *Edinburgh Review* 44 (1826): 315–41.

Macauley, James. Letter to Lyon Playfair, May 13, 1875. Imperial College of Science and Technology, Playfair Papers no. 469.

Maerker, Anna. *Model Experts: Wax Anatomies and Enlightenment in Florence and Vienna, 1775–1815*. Manchester: Manchester University Press, 2011.

Magee, Reginald. "Surgery in the Pre-Anaesthetic Era: The Life and Work of Robert Liston." *Australian and New Zealand Society of the History of Medicine* 2, no. 1 (2000): 121–33.

Magendie, François. "Expériences sur les fonctions des racines des nerfs qui naissent de la moëlle épinière." *Journal de physiologie expérimentale et de pathologie* 2 (1822): 366–71.

———. "Expériences sur les fonctions des racines des nerfs rachidiens." *Journal de physiologie expérimentale et de pathologie* 2 (1822): 276–79.

Marks, Harry M. *The Progress of Experiment: Science and Therapeutic Reform in the United States, 1900–1990*. Cambridge History of Medicine. Cambridge: Cambridge University Press, 1997.

Maulitz, Russell. "Channel Crossing: The Lure of French Pathology for English Medical Students, 1816–1836." *Bulletin of the History of Medicine* 55 (1981): 475–96.

———. *Morbid Appearances: The Anatomy of Pathology in the Early Nineteenth Century*. Cambridge History of Medicine. Cambridge: Cambridge University Press, 1987.

Mayo, Charles. *A Memoir of Pestalozzi: Being the Substance of a Lecture Delivered at the Royal Institution, Albemarle Street, May, 1826*. London: J. A. Hessey, 1828.

Mayo, Charles Herbert. *A Genealogical Account of the Mayo and Elton Families of the Counties of Wilts and Hereford*. London: C. Wittingham, 1882.

Mayo, Herbert. "Dilated Oesophagus." *London Medical Gazette* 3 (1828): 123–25.

———. "On Bellingeri's Claims as a Physiologist." *London Medical Gazette* 15 (1834): 271–72.

———. "On the Uses of the Facial Branches of the 5th and 7th Nerves." *London Medical Gazette* 3 (1829): 831–33.

———. *Outlines of Human Physiology*. 1st ed. London: Burgess and Hill, 1827.

———. *Outlines of Human Physiology*. 3rd ed. London: Burgess and Hill, 1833.

———. "To the Editor of the *Medical Quarterly Review*." *Medical Quarterly Review* 2 (1834): 450–51.

Mayo, Herbert, and Johann Christian Reil. *Anatomical and Physiological Commentaries*. 2 vols. London: printed for Thomas and George Underwood, 1822.

Mazumdar, Pauline. "Anatomical Physiology and the Reform of Medical Education." *Bulletin of the History of Medicine* 55 (1983): 475–96.

———. "Anatomy, Physiology and Surgery: Physiology Teaching in Early Nineteenth-Century London." *Canadian Bulletin of Medical History* 4, no. 2 (1987): 119–43.

Medical Times. "Professional Sketches." *Medical Times* 3, no. 57 (October 24 1840): 37.

Merton, Robert K. "Priorities in Scientific Discover: A Chapter in the Sociology of Science." *American Sociological Review* 22, no. 6 (1957): 635–59.

Messbarger, Rebecca Marie. *The Lady Anatomist: The Life and Work of Anna Morandi Manzolini*. Chicago: University of Chicago Press, 2010.

Miller, David Philip. *Discovering Water: James Watt, Henry Cavendish and the Nineteenth Century "Water Controversy."* Aldershot: Ashgate, 2004.

Morrell, Jack, and Arnold Thackray. *Gentlemen of Science: Early Years of the British Association for the Advancement of Science*. Oxford: Oxford University Press, 1981.

Morus, Iwan Rhys. *Frankenstein's Children: Electricity, Exhibition, and Experiment in Early-Nineteenth-Century London*. Princeton: Princeton University Press, 1998.

———. "Radicals, Romantics and Electrical Showmen: Placing Galvanism at the End of the English Enlightenment." *Notes and Records of the Royal Society* 63 (2009): 263–75.

Moseley, Maboth. *Irascible Genius: The Life of Charles Babbage*. Chicago: H. Regnery, 1970.

Mount, Harry. "Van Rymsdyk and the Nature-Menders: An Early Victim of the Two Cultures Divide." *British Journal for Eighteenth-Century Studies* 26 (2006): 79–96.

Newman, Charles. *The Evolution of Medical Education in the Nineteenth Century*. London: Oxford University Press, 1957.

Olmsted, J. M. D. *François Magendie, Pioneer in Experimental Physiology and Scientific Medicine in Nineteenth Century France*. New York: Schuman's, 1944.

Outram, Dorinda. *Georges Cuvier: Vocation, Science, and Authority in Post-Revolutionary France*. Manchester: Manchester University Press, 1984.

Paley, William. *Natural Theology*. London: Printed for R. Faulder by Wilks and Taylor, 1802.

Patrizio, Andrew, and Dawn Kemp. *Anatomy Acts: How We Come to Know Ourselves*. Edinburgh: Birlinn, 2006.

Petherbridge, Deanna, and L. J. Jordanova. *The Quick and the Dead: Artists and Anatomy.* Berkeley: University of California Press, 1997.

Philosophical Magazine: Comprehending the Various Branches of Science, Liberal and Fine Arts, Agriculture, Manufactures, and Commerce. "XLIX. Proceedings of Learned Societies: The Academy of Sciences at Berlin." *Philosophical Magazine* 18 (1804): 281.

Pichot, Amédée. *The Life and Labours of Sir Charles Bell.* London: R. Bentley, 1860.

Porter, Roy. *Disease, Medicine, and Society in England, 1550–1860.* Studies in Economic and Social History. Basingstoke: Macmillan Education, 1987.

———. *The Greatest Benefit to Mankind: A Medical History of Humanity.* New York: W. W. Norton, 1998.

———. "Medical Lecturing in Georgian London." *British Journal for the History of Science* 28, no. 1 (1995): 91–99.

———. "William Hunter, Surgeon." *History Today* 33, no. 9 (1983): 50–52.

Rauch, Alan. *Useful Knowledge: The Victorians, Morality, and the March of Intellect.* Durham: Duke University Press, 2001.

Rehbock, Philip. "Transcendental Anatomy." In *Romanticism and the Sciences,* edited by Andrew Cunningham and Nicholas Jardine, 144–61. Cambridge: Cambridge University Press, 1990.

Reinarz, Jonathan. "The Age of Museum Medicine: The Rise and Fall of the Medical Museum of Birmingham's School of Medicine." *Social History of Medicine* 18, no. 3 (2005): 419–46.

Report of the Royal Commission on the Practice of Subjecting Live Animals to Experiments for Scientific Purposes. London: George Edward Eyre and William Spottiswoode, 1876.

Rice, Gillian. "The Bell-Magendie-Walker Controversy." *Medical History* 31, no. 2 (1987): 190–200.

Richards, Robert J. *The Meaning of Evolution: The Morphological Construction and Ideological Reconstruction of Darwin's Theory.* Science and Its Conceptual Foundations. Chicago: University of Chicago Press, 1992.

Richardson, Alan. *British Romanticism and the Science of the Mind.* Cambridge Studies in Romanticism. Cambridge: Cambridge University Press, 2001.

Richardson, Ruth. *Death, Dissection, and the Destitute.* 2nd ed. Chicago: University of Chicago Press, 2001.

Roberts, Lissa. "The Death of the Sensuous Chemist." *Studies in the History and Philosophy of Science* 26 (1995): 503–29.

Romano, Terrie M. *Making Medicine Scientific: John Burdon Sanderson and the Culture of Victorian Science.* Baltimore: Johns Hopkins University Press, 2002.

Rosner, Lisa. *Medical Education in the Age of Improvement: Edinburgh Students and Apprentices 1760–1826.* Edinburgh: Edinburgh University Press, 1991.

Rupke, Nicolaas A. *Vivisection in Historical Perspective.* London: Croom Helm, 1987.

Ruskin, Steven. *John Herschel's Cape Voyage: Private Science, Public Imagination, and the Ambitions of Empire.* Burlington: Ashgate, 2004.

Russell, E. S. *Form and Function: A Contribution to the History of Animal Morphology.* Chicago: University of Chicago Press, 1982.

Sappol, Michael. *Dream Anatomy.* NIH Publication. Washington, DC: US Dept. of Health and Human Services, National Institutes of Health, National Library of Medicine, 2006.

———. *A Traffic of Dead Bodies: Anatomy and Embodied Social Identity in Nineteenth-Century America.* Princeton: Princeton University Press, 2002.

Sarafianos, Aris. "B. R. Haydon and Racial Science: The Politics of the Human Figure and the Art Profession in the Early Nineteenth Century." *Visual Culture in Britain* 7, no. 1 (2006): 79–106.

Secord, James A. *Victorian Sensation: The Extraordinary Publication, Reception, and Secret Authorship of Vestiges of the Natural History of Creation.* Chicago: University of Chicago Press, 2000.

Shapin, Steven. "The Audience for Science in Eighteenth Century Edinburgh." *History of Science* (1974): 95–121.

———. "Of Gods and Kings: Natural Philosophy and Politics in the Leibniz-Clarke Disputes." *Isis* 72, no. 2 (1981): 187–215.

———. "Phrenological Knowledge and the Social Structure of Early Nineteenth-Century Edinburgh." *Annals of Science* 32, no. 3 (1975): 219–43.

———. "The Politics of Observation: Cerebral Anatomy and Social Interests in the Edinburgh Phrenology Disputes." In *On the Margins of Science: The Social Construction of Rejected Knowledge*, edited by Roy Wallis, 139–78. Sociological Review Monographs 27. Keele: University of Keele, 1979.

Shapin, Steven, and Barry Barnes. "Head and Hand: Rhetorical Resources in British Pedagogical Writing, 1770–1850." *Oxford Review of Education* 2, no. 3 (1976): 231–54.

Shapin, Steven, and Adi Ophir. "The Place of Knowledge: A Methodological Survey." *Science in Context* 4, no. 1 (1991): 3–21.

Shaw, Alexander. *An Account of Discoveries of Sir Charles Bell in the Nervous System*. London: J. Murray, 1860.

———. "Letter to the Editor, Mr. Shaw in Reply to Mr. Mayo." *London Medical Gazette* 4 (1829): 12–14.

———. *Narrative of the Discoveries of Sir Charles Bell in the Nervous System*. London: Longman, 1839.

Shortt, Samuel. "Physicians, Science, and Status: Issues in the Professionalization of Anglo-American Medicine in the Nineteenth Century." *Medical History* 27 (1983): 51–68.

Singer, Charles, and S. W. F. Holloway. "Early Medical Education in England in Relation to the Pre-History of London University." *Medical History* 4, no. 1 (1960): 1–17.

Smith, Adam. *The Theory of Moral Sentiments*. London: A. Millar, 1759.

Smith, Adam, and James R. Otteson. *Adam Smith: Selected Philosophical Writings*. Library of Scottish Philosophy. Exeter: Imprint Academic, 2004.

Sprigge, Samuel Squire. *The Life and Times of Thomas Wakley*. London: Longman, Green and Co., 1897.

Statement by the Council of the University of London, Explanatory of the Nature and Objects of the Institution. London: Longman, Rees, Orme, Brown, and Green; and John Murray, Albemarle Street, 1827.

Stewart, Dugald. "Account of the Life and Writings of Adam Smith." In *The Collected Works of Dugald Stewart*, edited by Bart. Sir William Hamilton, vol. X, 5–100. Edinburgh: Thomas Constable and Co, 1858.

———. *Elements of the Philosophy of the Human Mind*. In *The Collected Works of Dugald Stewart*, edited by William Hamilton and John Veitch. Edinburgh: Thomas Constable and Co., [1827] 1854.

Struthers, John. *Historical Sketch of the Edinburgh Anatomy School*. Edinburgh: Maclachlan and Stewart, 1867.

Swade, Doron, and Charles Babbage. *The Difference Engine: Charles Babbage and the Quest to Build the First Computer*. New York: Viking, 2001.

Tannoch-Bland, Jennifer. "Dugald Stewart on Intellectual Character." *British Journal for the History of Science* 30 (1997): 307–20.

Thomson, John. *An Account of the Life, Lectures, and Writing of William Cullen*. 2 vols. Edinburgh: Blackwood, 1859.

Thomson, Stewart Craig. "The Surgeon-Anatomists of Great Windmill Street School." *Bulletin of the Society of Medical History of Chicago* 5 (1937–46): 301–21.

Thornton, John Leonard. *John Abernethy: A Biography*. London: printed for the author; distributed by Simpkin Marshall, 1953.

Topham, Jonathan. "Beyond the 'Common Context': The Production and Reading of the Bridgewater Treatises." *Isis* 89 (1998): 233–62.

———. "Science and Popular Education in the 1830s: The Role of the 'Bridgewater Treatises.'" *British Journal for the History of Science* 25, no. 4 (1992): 397–430.

———. "Scientific Publishing and the Reading of Science in Nineteenth-Century Britain: A Historiographical Survey and Guide to Sources." *Studies in the History and Philosophy of Science* 31 (2000): 559–612.

Van Wyhe, John. *Phrenology and the Origins of Victorian Scientific Naturalism*. Aldershot: Ashgate, 2004.

Waddington, Keir. *Medical Education at St. Bartholomew's Hospital, 1123–1995*. Woodbridge: Boydell Press, 2003.

Wakley, Thomas. "Editorial." *Lancet* 1829–30, vol. 1 (1829): 42–49.

———. "The London College of Medicine." *Lancet* 1830–31, vol. 2 (1831): 243–50.

———. "London College of Medicine: June 18, 1831." *Lancet* 1830–31, vol. 2 (1831): 379–80.

———. "London College of Medicine: May 7, 1831." *Lancet* 1830–31, vol. 2 (1831): 177–83.

———. "Reply to the Slanderers." *Lancet* 1829–30, vol. 1 (1829): 1–5.

Warner, John Harley. *Against the Spirit of System: The French Impulse in Nineteenth-Century American Medicine*. Princeton: Princeton University Press, 1998.

———. "The History of Science and the Sciences of Medicine." *Osiris* 10 (1995): 164–93.

———. "The Idea of Science in English Medicine: The 'Decline of Science' and the Rhetoric of Reform, 1815–1845." In *British Medicine in an Age of Reform*, edited by Roger French and Andrew Wear. London: Routledge, 1991.

Wear, A. *Medicine in Society: Historical Essays*. Cambridge: Cambridge University Press, 1992.

Wilson, William James Erasmus. *The History of the Middlesex Hospital During the First Century of Its Existence*. Oxford: Oxford University, 1843.

Index

Page numbers in italics refer to illustrations.